COMMUNICATION SYSTEMS
The State of the Art

IFIP - The International Federation for Information Processing

IFIP was founded in 1960 under the auspices of UNESCO, following the First World Computer Congress held in Paris the previous year. An umbrella organization for societies working in information processing, IFIP's aim is two-fold: to support information processing within its member countries and to encourage technology transfer to developing nations. As its mission statement clearly states,

IFIP's mission is to be the leading, truly international, apolitical organization which encourages and assists in the development, exploitation and application of information technology for the benefit of all people.

IFIP is a non-profitmaking organization, run almost solely by 2500 volunteers. It operates through a number of technical committees, which organize events and publications. IFIP's events range from an international congress to local seminars, but the most important are:

- The IFIP World Computer Congress, held every second year;
- open conferences;
- working conferences.

The flagship event is the IFIP World Computer Congress, at which both invited and contributed papers are presented. Contributed papers are rigorously refereed and the rejection rate is high.

As with the Congress, participation in the open conferences is open to all and papers may be invited or submitted. Again, submitted papers are stringently refereed.

The working conferences are structured differently. They are usually run by a working group and attendance is small and by invitation only. Their purpose is to create an atmosphere conducive to innovation and development. Refereeing is less rigorous and papers are subjected to extensive group discussion.

Publications arising from IFIP events vary. The papers presented at the IFIP World Computer Congress and at open conferences are published as conference proceedings, while the results of the working conferences are often published as collections of selected and edited papers.

Any national society whose primary activity is in information may apply to become a full member of IFIP, although full membership is restricted to one society per country. Full members are entitled to vote at the annual General Assembly, National societies preferring a less committed involvement may apply for associate or corresponding membership. Associate members enjoy the same benefits as full members, but without voting rights. Corresponding members are not represented in IFIP bodies. Affiliated membership is open to non-national societies, and individual and honorary membership schemes are also offered.

COMMUNICATION SYSTEMS

The State of the Art

IFIP 17th World Computer Congress —
TC6 Stream on Communication Systems: The State of the Art
August 25-30, 2002, Montréal, Québec, Canada

Edited by

Lyman Chapin
NextHop Technologies
USA

KLUWER ACADEMIC PUBLISHERS
BOSTON / DORDRECHT / LONDON

Distributors for North, Central and South America:
Kluwer Academic Publishers
101 Philip Drive
Assinippi Park
Norwell, Massachusetts 02061 USA
Telephone (781) 871-6600
Fax (781) 681-9045
E-Mail <kluwer@wkap.com>

Distributors for all other countries:
Kluwer Academic Publishers Group
Post Office Box 322
3300 AH Dordrecht, THE NETHERLANDS
Telephone 31 786 576 000
Fax 31 786 576 474
E-Mail <services@wkap.nl>

 Electronic Services <http://www.wkap.nl>

Library of Congress Cataloging-in-Publication Data

A C.I.P. Catalogue record for this book is available from the Library of Congress.

Communication Systems: The State of the Art
Edited by Lyman Chapin
ISBN 1-4020-7168-X

Copyright © 2002 by International Federation for Information Processing.

All rights reserved. No part of this work may be reproduced, stored in a retrieval system, or transmitted in any form or by any means, electronic, mechanical, photocopying, microfilming, recording, or otherwise, without written permission from the Publisher (Kluwer Academic Publishers, 101 Philip Drive, Assinippi Park, Norwell, Massachusetts 02061), with the exception of any material supplied specifically for the purpose of being entered and executed on a computer system, for exclusive use by the purchaser of the work.

Printed on acid-free paper.
Printed in the United States of America.

Contents

Preface vii

Research Advances in Middleware for Distributed Systems:
State of the Art
    Richard E. Schantz, Douglas C. Schmidt 1

Internet Routing: The State of the Art
    K.Ramasamy, A.Arokiasamy, P.A.Balakrishnan 37

Performance of Telecommunication Systems: State of the Art
    *Kimon Kontovasilis, S. Wittevrongel, H. Bruneel, B. Van Houdt,
    Chris Blondia* 61

Enhancing Web Performance
    Arun Iyengar, Erich Nahum, Anees Shaikh, Renu Tewari 95

Network Management: State of the Art
    Raouf Boutaba, Jin Xiao 127

State of the Art of Service Creation Technologies in IP and Mobile
Environments
    Jorma Jormakka, Henryka Jormakka 147

QoS, Security, and Mobility Management for Fixed and Wireless
Networks under Policy-based Techniques
    Guy Pujolle, Hakima Chaouchi 167

A Multicast Routing Protocol with Multiple QoS Constraints
  Li Layuan, Li Chunlin 181

Anonymous Internet Communication Based on IPSec
  Ronggong Song, Larry Korba 199

Internet Interconnection Economic Model and its Analysis:
Peering and Settlement
  Martin B. Weiss, Seung Jae Shin 215

Dimensioning Company Intranets for IT Bandwidth
  Sandor Vincze 233

Preface

Until recently, communication systems played a minor supporting role in the evolution of computing from mainframes to minis to PCs and servers; today, they are an integral part of almost every aspect of computing, from distributed applications to the Internet. Even experts have trouble keeping up with all of the latest developments in a field that is growing and changing so rapidly.

Recognizing the need for researchers and practitioners in all information technology fields to keep up with the state of the art in communications, IFIP's Technical Committee on Communication Systems, TC6, assembled a world-class roster of individual area experts to present a comprehensive "State of the Art in Communication Systems" programme at the 17th World Computer Congress, which was sponsored by the International Federation for Information processing (IFIP) and held in Montréal, Québec, Canada in August 2002.

The papers collected in this book capture the depth and breadth of the field of communication systems, from the basic unicast and multicast routing infrastructure of modern IP-centered data networks, to the middleware services that facilitate the construction of useful distributed systems, to the applications—especially those that take advantage of the world-wide web—that ultimately deliver value to users. The many corollary dimensions of real-world networking are represented here as well, including communication system performance, network management, security, Internet economics, and wireless mobility.

In addition to the papers collected in this book, the following two invited talks in the communication systems stream were presented during WCC 2002:

- André Danthine, Université de Liège, Belgium presented "A New Internet Architecture with MPLS and GMPLS."
- Gerard J. Holzmann, Bell Laboratories, USA presented "Static and Dynamic Software Testing: State of the Art."

Lyman Chapin

Preface

Until recently, communication systems played a minor supporting role in the evolution of computing from mainframes to minis to PCs and servers; today, they are an integral part of almost every aspect of computing, from distributed applications to the Internet. Even experts have trouble keeping up with all of the latest developments in a field that is growing and changing so rapidly.

Recognizing the need for researchers and practitioners in all information technology fields to keep up with the state of the art in communications, IFIP's Technical Committee on Communication Systems, TC6, assembled a world-class roster of individual area experts to present a comprehensive "State of the Art in Communication Systems" programme at the 17th World Computer Congress, which was sponsored by the International Federation for Information processing (IFIP) and held in Montréal, Québec, Canada in August 2002.

The papers collected in this book capture the depth and breadth of the field of communication systems, from the basic unicast and multicast routing infrastructure of modern IP-centered data networks, to the middleware services that facilitate the construction of useful distributed systems, to the applications—especially those that take advantage of the world-wide web—that ultimately deliver value to users. The many corollary dimensions of real-world networking are represented here as well, including communication system performance, network management, security, Internet economics, and wireless mobility.

In addition to the papers collected in this book, the following two invited talks in the communication systems stream were presented during WCC 2002:

- André Danthine, Université de Liège, Belgium presented "A New Internet Architecture with MPLS and GMPLS."
- Gerard J. Holzmann, Bell Laboratories, USA presented "Static and Dynamic Software Testing: State of the Art."

Lyman Chapin

Research Advances in Middleware for Distributed Systems: State of the Art

Richard E. Schantz
BBN Technologies
schantz@bbn.com

Douglas C. Schmidt
Electrical & Computer Engineering Dept.
University of California, Irvine
schmidt@uci.edu

1 INTRODUCTION

Two fundamental trends influence the way we conceive and construct new computing and information systems. The first is that *information technology of all forms is becoming highly commoditized i.e.,* hardware and software artifacts are getting faster, cheaper, and better at a relatively predictable rate. The second is the *growing acceptance of a network-centric paradigm*, where distributed applications with a range of quality of service (QoS) needs are constructed by integrating separate components connected by various forms of communication services. The nature of these interconnections can range from very small and tightly coupled systems, such as avionics mission computing systems, to very large and loosely coupled systems, such as global telecommunications systems and so-called "grid" computing.

The interplay of these two trends has yielded new architectural concepts and services embodying layers of *middleware*. Middleware is systems software that resides between the applications and the underlying operating systems, network protocol stacks, and hardware. Its primary role is to functionally bridge the gap between application programs and the lower-level hardware and software infrastructure in order to:

1. Make it feasible, easier, and more cost effective to develop and evolve distributed systems
2. Coordinate how parts of applications are connected and how they interoperate and
3. Enable and simplify the integration of components developed by multiple technology suppliers.

The growing importance of middleware stems from the recognition of the need for more advanced and capable support–beyond simple connectivity–to construct effective distributed systems. A significant portion of middleware-oriented R&D activities over the past decade have therefore focused on

1. Identifying, evolving, and expanding our understanding of current middleware services to support the network-centric paradigm and
2. Defining additional middleware layers and capabilities to meet the challenges associated with constructing future network-centric systems.

These activities are expected to continue forward well into this decade to address the needs of next-generation distributed systems.

The past decade has yielded significant progress in middleware, which has stemmed in large part from the following trends:

- *Years of iteration, refinement, and successful use* – The use of middleware and middleware oriented system architectures is not new [Sch86, Sch98, Ber96]. Middleware concepts emerged alongside experimentation with the early Internet (and even its predecessor the ARPAnet) and systems based on middleware have been operational continuously since the mid 1980's. Over that period of time, the ideas, designs, and (most importantly) the software that incarnates those ideas have had a chance to be tried and refined (for those that worked), and discarded or redirected (for those that didn't). This iterative technology development process takes a good deal of time to get right and be accepted by user communities, and a good deal of patience to stay the course. When this process is successful, it often results in frameworks, components, and patterns that reify the knowledge of how to apply these technologies, along with standards that codify the boundaries of these technologies, as described next.
- *The dissemination of middleware frameworks, components, and patterns* – During the past decade, a substantial amount of R&D effort

has focused on developing *frameworks, components,* and *patterns* as a means to promote the development and reuse of successful middleware technology. Patterns capture successful solutions to commonly occurring software problems that arise in a particular context [Gam95]. Patterns can simplify the design, construction, and performance tuning of middleware and applications by codifying the accumulated expertise of developers who have confronted similar problems before. Patterns also raise the level of discourse in describing software design and programming activities. Frameworks and components are concrete realizations of groups of related patterns [John97]. Well-designed frameworks reify patterns in terms of functionality provided by components in the middleware itself, as well as functionality provided by an application. Frameworks also integrate various approaches to problems where there are no *a priori*, context-independent, optimal solutions. Middleware frameworks [Sch02] and component toolkits can include strategized selection and optimization patterns so that multiple independently developed capabilities can be integrated and configured automatically to meet the functional and QoS requirements of particular applications.

- ***The maturation of middleware standards*** – Also over the past decade, standards for middleware frameworks and components have been established and have matured considerably. For instance, the Object Management Group (OMG) has defined the following specifications for CORBA [Omg00] in the past several years:

  o *CORBA Component Model*, which standardizes component implementation, packaging, and deployment to simplify server programming and configuration
  o *Minimum CORBA*, which removes non-essential features from the full OMG CORBA specification to reduce footprint so that CORBA can be used in memory-constrained embedded systems
  o *Real-time CORBA*, which includes features that allow applications to reserve and manage network, CPU, and memory resources predictably end-to-end
  o *CORBA Messaging*, which exports additional QoS policies, such as timeouts, request priorities, and queueing disciplines, to applications and
  o *Fault-tolerant CORBA*, which uses entity redundancy of objects to support replication, fault detection, and failure recovery.

These middleware specifications go well beyond the interconnection standards conceived originally and reflect attention to more detailed issues beyond basic connectivity and remote procedure calls. Robust implementations of these CORBA capabilities and services are now available from multiple suppliers, some using open-source business models and some using traditional business models.

In addition to CORBA, some other notable successes to date in the domain of middleware services, frameworks, components, and standards include:

- Java 2 Enterprise Edition (J2EE) [Tho98] and .NET *[NET01]*, which have introduced advanced software engineering capabilities to the mainstream IT community and which incorporate various levels of middleware as part of the overall development process, albeit with only partial support for performance critical and embedded solutions.
- Akamai et al, which have legitimized a form of middleware service as a viable business, albeit using proprietary and closed, non-user programmable solutions.
- Napster, which demonstrated the power of having a powerful, commercial-off-the-shelf (COTS) middleware infrastructure to start from in quickly (weeks/months) developing a very capable system, albeit without much concern for system lifecycle and software engineering practices, i.e., it is one of a kind.
- WWW, where the world wide web middleware/standards led to easily connecting independently developed browsers and web pages, albeit also the world wide *wait*, because there was no system engineering or attention paid to enforcing end-to-end quality of service issues.
- The Global Grid, which is enabling scientists and high performance computing researchers to collaborate on grand challenge problems, such as global climate change modeling, albeit using architectures and tools that are not yet aligned with mainstream IT COTS middleware.

These competing and complementary forms of and approaches to middleware based solutions represent simultaneously a healthy and robust technical area of continuing innovation, and a source of confusion due to the multiple forms of similar capabilities, patterns, and architectures.

2 ADDRESSING DISTRIBUTED APPLICATION CHALLENGES WITH MIDDLEWARE

Requirements for faster development cycles, decreased effort, and greater software reuse motivate the creation and use of middleware and middleware-based architectures. When implemented properly, middleware helps to:

- Shield software developers from low-level, tedious, and error-prone platform details, such as socket-level network programming.
- Amortize software lifecycle costs by leveraging previous development expertise and capturing implementations of key patterns in reusable frameworks, rather than rebuilding them manually for each use.
- Provide a consistent set of higher-level network-oriented abstractions that are much closer to application requirements in order to simplify the development of distributed and embedded systems.
- Provide a wide array of reuseable, off-the-shelf developer-oriented services, such as naming, logging and security that have proven necessary to operate effectively in a networked environment.

Over the past decade, various middleware technologies have been devised to alleviate many complexities associated with developing software for distributed applications. Their successes have added middleware as a new category of systems software to complement the familiar operating system, programming language, networking, and database offerings of the previous generation. Some of the most successful middleware technologies have centered on *distributed object computing (DOC)*. DOC is an advanced, mature, and field-tested middleware connectivity paradigm that also supports flexible and adaptive behavior. DOC middleware architectures are composed of relatively autonomous software objects that can be distributed or collocated throughout a wide range of networks and interconnects. Clients invoke operations on target objects to perform interactions and invoke functionality needed to achieve application goals. Through these interactions, a wide variety of middleware-based services are made available off-the-shelf to simplify application development. Aggregations of these simple, middleware-mediated interactions are increasingly forming the basis of large-scale distributed system deployments.

2.1 The Structure and Functionality of DOC Middleware

Just as networking protocol stacks can be decomposed into multiple layers, such as the physical, data-link, network, transport, session, presentation, and application layers, so too can DOC middleware be decomposed into multiple layers, such as those shown in Figure 1.

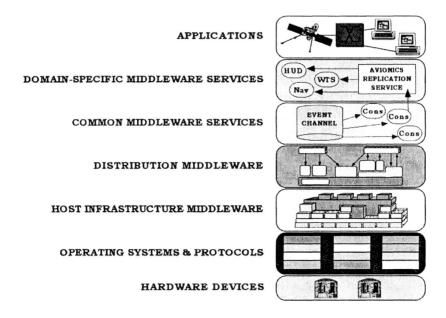

Figure 1. Layers of DOC Middleware and Surrounding Context

Below, we describe each of these middleware layers and outline some of the COTS technologies in each layer that have matured and found widespread use in recent years.

2.1.1 Host infrastructure middleware

Host infrastructure middleware encapsulates and enhances native OS communication and concurrency mechanisms to create reusable network programming components, such as reactors, acceptor-connectors, monitor objects, active objects, and component configurators [Sch00b]. These components abstract away the peculiarities of individual operating systems and help eliminate many tedious, error-prone, and non-portable aspects of developing and maintaining networked applications via low-level OS programming APIs, such as Sockets or POSIX pthreads. Widely used examples of host infrastructure middleware include:

- *The Sun Java Virtual Machine (JVM) [JVM97], which* provides a platform-independent way of executing code by abstracting the differences between operating systems and CPU architectures. A JVM is responsible for interpreting Java bytecode, and for translating the bytecode into an action or operating system call. It is the JVM's responsibility to encapsulate platform details within the portable

bytecode interface, so that applications are shielded from disparate operating systems and CPU architectures on which Java software runs.
- *.NET [NET01]* is Microsoft's platform for XML Web services, which are designed to connect information, devices, and people in a common, yet customizable way. The common language runtime (CLR) is the host infrastructure middleware foundation upon which Microsoft's .NET services are built. The Microsoft CLR is similar to Sun's JVM, *i.e.,* it provides an execution environment that manages running code and simplifies software development via automatic memory management mechanisms, cross-language integration, interoperability with existing code and systems, simplified deployment, and a security system.
- The ADAPTIVE Communication Environment (ACE) is a freely available, highly portable toolkit that shields applications from differences between native OS programming capabilities, such as file handling, connection establishment, event demultiplexing, interprocess communication, (de)marshaling, concurrency, and synchronization. ACE provides an OS adaptation layer and wrapper facades [Sch02] that encapsulate OS file system, concurrency, and network programming mechanisms. ACE also provides reusable frameworks [Sch03] that handle network programming tasks, such as synchronous and asynchronous event handling, service configuration and initialization, concurrency control, connection management, and hierarchical service integration.

The primary differences between ACE, JVMs, and the .NET CLR are that (1) ACE is always a compiled C++ interface, rather than an interpreted bytecode interface, which removes a level of indirection and helps to optimize runtime performance, (2) ACE is open-source, so it's possible to subset it or modify it to meet a wide variety of needs, and (3) ACE runs on more OS and hardware platforms than JVMs and CLR.

2.1.2 Distribution middleware

Distribution middleware defines higher-level distributed programming models whose reusable APIs and components automate and extend the native OS network programming capabilities encapsulated by host infrastructure middleware. Distribution middleware enables clients to program distributed applications much like stand-alone applications, i.e., by invoking operations on target objects without hard-coding dependencies on their location, programming language, OS platform, communication protocols and interconnects, and hardware. At the heart of distribution middleware are request brokers, such as:

- The OMG's Common Object Request Broker Architecture (CORBA) [Omg00], which allows objects to interoperate across networks regardless of the language in which they were written or the platform on which they are deployed.
- Sun's Java Remote Method Invocation (RMI) [Wol96], which enables developers to create distributed Java-to-Java applications, in which the methods of remote Java objects can be invoked from other JVMs, possibly on different hosts. RMI supports more sophisticated object interactions by using object serialization to marshal and unmarshal parameters, as well as whole objects. This flexibility is made possible by Java's virtual machine architecture and is greatly simplified by using a single language..
- Microsoft's Distributed Component Object Model (DCOM) [Box97], which enables software components to communicate over a network via remote component instantiation and method invocations. Unlike CORBA and Java RMI, which run on many operating systems, DCOM is implemented primarily on Windows platforms.
- SOAP [SOAP01] is an emerging distribution middleware technology based on a lightweight and simple XML-based protocol that allows applications to exchange structured and typed information on the Web. SOAP is designed to enable automated Web services based on a shared and open Web infrastructure. SOAP applications can be written in a wide range of programming languages, used in combination with a variety of Internet protocols and formats (such as HTTP, SMTP, and MIME), and can support a wide range of applications from messaging systems to RPC.

An example of distribution middleware R&D is the TAO project [Sch98a] conducted by researchers at Washington University, St. Louis, the University of California, Irvine, and Vanderbilt University as part several DARPA programs. TAO is an open-source Real-time CORBA ORB that allows distributed real-time and embedded (DRE) applications to reserve and manage

- *Processor resources* via thread pools, priority mechanisms, intra-process mutual exclusion mechanisms, and a global scheduling service for real-time systems with fixed priorities
- *Communication resources* via protocol properties and explicit bindings to server objects using priority bands and private connections and
- *Memory resources* via buffering requests in queues and bounding the size of thread pools.

TAO is implemented with reusable frameworks from the ACE [Sch02, Sch03] host infrastructure middleware toolkit. ACE and TAO are mature examples of middleware R&D transition, having been used in hundreds of DRE systems, including telecom network management and call processing, online trading services, avionics mission computing, software defined radios, radar systems, surface mount "pick and place" systems, and hot rolling mills.

2.1.3 Common middleware services

Common middleware services augment distribution middleware by defining higher-level domain-independent services that allow application developers to concentrate on programming business logic, without the need to write the "plumbing" code required to develop distributed applications by using lower-level middleware directly. For example, application developers no longer need to write code that handles transactional behavior, security, database connection pooling or threading, because common middleware service providers bundle these tasks into reusable components. Whereas distribution middleware focuses largely on managing end-system resources in support of an object-oriented distributed programming model, common middleware services focus on allocating, scheduling, and coordinating various resources throughout a distributed system using a component programming and scripting model. Developers can reuse these component services to manage global resources and perform common distribution tasks that would otherwise be implemented in an ad hoc manner within each application. The form and content of these services will continue to evolve as the requirements on the applications being constructed expand. Examples of common middleware services include:

- The OMG's CORBA Common Object Services (CORBAservices) [Omg98b], which provide domain-independent interfaces and capabilities that can be used by many DOC applications. The OMG CORBAservices specifications define a wide variety of these services, including event notification, logging, multimedia streaming, persistence, security, global time, real-time scheduling, fault tolerance, concurrency control, and transactions.
- Sun's Java 2 Enterprise Edition (J2EE) technology [Tho98], which allows developers to create n-tier distributed systems by linking a number of pre-built software services—called "Javabeans"—without having to write much code from scratch. Since J2EE is built on top of Java technology, J2EE service components can only be implemented using the Java language. The CORBA Component Model (CCM) [Omg99] defines a superset of J2EE capabilities that can be

implemented using all the programming languages supported by CORBA.

- Microsoft's .NET Web services [NET01], which complements the lower-level middleware .NET capabilities, allows developers to package application logic into components that are accessed using standard higher-level Internet protocols above the transport layer, such as HTTP. The .NET Web services combine aspects of component-based development and Web technologies. Like components, .NET Web services provide black-box functionality that can be described and reused without concern for how a service is implemented. Unlike traditional component technologies, however, .NET Web services are not accessed using the object model–specific protocols defined by DCOM, Java RMI, or CORBA. Instead, XML Web services are accessed using Web protocols and data formats, such as the Hypertext Transfer Protocol (HTTP) and eXtensible Markup Language (XML), respectively.

2.1.4 Domain-specific middleware services

Domain-specific middleware services are tailored to the requirements of particular domains, such as telecom, e-commerce, health care, process automation, or aerospace. Unlike the other three DOC middleware layers, which provide broadly reusable "horizontal" mechanisms and services, domain-specific middleware services are targeted at vertical markets. From a COTS perspective, domain-specific services are the least mature of the middleware layers today. This immaturity is due partly to the historical lack of distribution middleware and common middleware service standards, which are needed to provide a stable base upon which to create domain-specific services. Since they embody knowledge of a domain, however, domain-specific middleware services have the most potential to increase system quality and decrease the cycle-time and effort required to develop particular types of networked applications. Examples of domain-specific middleware services include the following:

- The OMG has convened a number of Domain Task Forces that concentrate on standardizing domain-specific middleware services. These task forces vary from the Electronic Commerce Domain Task Force, whose charter is to define and promote the specification of OMG distributed object technologies for the development and use of Electronic Commerce and Electronic Market systems, to the Life Science Research Domain Task Force, who do similar work in the area of Life Science, maturing the OMG specifications to improve the quality and utility of software and information systems used in Life Sciences

Research. There are also OMG Domain Task Forces for the healthcare, telecom, command and control, and process automation domains.
- The Siemens Medical Engineering Group has developed Syngo(R), which is both an integrated collection of domain-specific middleware services, as well as an open and dynamically extensible application server platform for medical imaging tasks and applications, including ultrasound, mammography, radiography, flouroscopy, angiography, computer tomography, magnetic resonance, nuclear medicine, therapy systems, cardiac systems, patient monitoring systems, life support systems, and imaging- and diagnostic-workstations. The Syngo(R) middleware services allow healthcare facilities to integrate diagnostic imaging and other radiological, cardiological and hospital services via a blackbox application template framework based on advanced patterns for communication, concurrency, and configuration for both business logic and presentation logic supporting a common look and feel throughout the medical domain.
- The Boeing Bold Stroke [Sha98, Doe99] architecture uses COTS hardware and middleware to produce a non-proprietary, standards-based component architecture for military avionics mission computing capabilities, such as navigation, display management, sensor management and situational awareness, data link management, and weapons control. A driving objective of Bold Stroke is to support reusable product line applications, leading to a highly configurable application component model and supporting middleware services. Associated products ranging from single processor systems with $O(10^5)$ lines of source code to multi-processor systems with $O(10^6)$ lines of code have shown dramatic affordability and schedule improvements and have been flight tested successfully. The domain-specific middleware services in Bold Stroke are layered upon common middleware services (the CORBA Event Service), distribution middleware (Real-time CORBA), and host infrastructure middleware (ACE), and have been demonstrated to be highly portable for different COTS operating systems (e.g. VxWorks), interconnects (e.g. VME), and processors (e.g. PowerPC).

2.2 The Benefits of DOC Middleware

Middleware in general–and DOC middleware in particular–provides essential capabilities for developing an increasingly large class of distributed applications. In this section we summarize some of the improvements and areas of focus in which middleware oriented approaches are having significant impact.

2.2.1 Growing focus on integration rather than on programming

This visible shift in focus is perhaps the major accomplishment of currently deployed middleware. Middleware originated because the problems relating to integration and construction by composing parts were not being met by either

- Applications, which at best were customized for a single use,
- Networks, which were necessarily concerned with providing the communication layer, or
- Host operating systems, which were focused primarily on a single, self-contained unit of resources.

In contrast, middleware has a fundamental integration focus, which stems from incorporating the perspectives of both operating systems and programming model concepts into organizing and controlling the composition of separately developed components across host boundaries. Every DOC middleware technology has within it some type of request broker functionality that initiates and manages inter-component interactions.

Distribution middleware, such as CORBA, Java RMI, or SOAP, makes it easy and straightforward to connect separate pieces of software together, largely independent of their location, connectivity mechanism, and technology used to develop them. These capabilities allow DOC middleware to amortize software life-cycle efforts by leveraging previous development expertise and reifying implementations of key patterns into more encompassing reusable frameworks and components. As DOC middleware continues to mature and incorporates additional needed services, next-generation applications will increasingly be assembled by modeling, integrating, and scripting domain-specific and common service components, rather than by being programmed either entirely from scratch or requiring significant customization or augmentation to off-the-shelf component implementations.

2.2.2 The increased viability of open systems architectures and open-source availability

By their very nature, systems developed by composing separate components are more open than systems conceived and developed as monolithic entities. The focus on interfaces for integrating and controlling the component parts leads naturally to *standard* interfaces. This in turn yields the potential for multiple choices for component implementations and to open engineering concepts. Standards organizations such as the OMG and The Open Group have fostered the cooperative efforts needed to bring together groups of users and vendors to define domain-specific functionality that overlays open integrating architectures, forming a basis for industry-wide use of some software components. Once a common, open structure exists, it becomes feasible for a wide variety of participants to contribute to the off-the-shelf availability of additional parts needed to construct complete systems. Since few companies today can afford significant investments in internally funded R&D, it is increasingly important for the information technology industry to leverage externally funded R&D sources, such as government investment. In this context, standards-based DOC middleware serves as a common platform to help concentrate the results of R&D efforts and ensure smooth transition conduits from research groups into production systems.

For example, research conducted under the DARPA Quorum program [Quorum99] focused heavily on CORBA open systems middleware. Quorum yielded many results that transitioned into standardized service definitions and implementations for Real-time [OMG00B, Sch98a] and Fault-tolerant [Omg98a, Cuk98] CORBA specification and productization efforts. In this case, focused government R&D efforts leveraged their results by exporting them into, and combining them with, other on going public and private activities that also used a standards-based open middleware substrate. Prior to the viability of common middleware platforms, these same results would have been buried within a custom or proprietary system, serving only as the existence proof, not as the basis for incorporating into a larger whole.

2.2.3 Increased leverage for disruptive technologies leading to increased global competition

Middleware that supports component integration and reuse is a key technology to help amortize software life-cycle costs by:

1. Leveraging previous development expertise, *e.g.*, DOC middleware helps to abstract commonly reused low-level OS concurrency and networking details away into higher-level, more easily used artifacts and
2. Focusing on efforts to improve software quality and performance, *e.g.*, DOC middleware combines various aspects of a larger solution together, *e.g.*, fault tolerance for domain-specific objects with real-time QoS properties.

When developers needn't worry as much about low-level details they are freed to focus on more strategic, larger scope, application-centric specializations concerns, such as distributed resource management and end-to-end dependability. Ultimately, this higher level focus will result in software-intensive distributed system components that apply reusable middleware to get smaller, faster, cheaper, and better at a predictable pace, just as computing and networking hardware do today. And that, in turn, will enable the next-generation of better and cheaper approaches to what are now carefully crafted custom solutions, which are often inflexible and proprietary. The result will be a new technological economy where developers can leverage frequently used common components, which come with steady innovation cycles resulting from a multi-user basis, in conjunction with custom domain-specific capabilities, which allow appropriate mixing of multi-user low cost and custom development for competitive advantage.

2.2.4 Growing focus on real-time embedded environments integrating computational and real world physical assets

Historically, conventional COTS software has been unsuitable for use in mission-critical distributed systems due to its either being flexible and standard, but incapable of guaranteeing stringent QoS demands (which restricts assurability) or partially QoS-enabled, but inflexible and non-standard (which restricts adaptability and affordability). As a result, the rapid progress in COTS software for mainstream desktop business information technology (IT) has not yet become as broadly applicable for mission-critical distributed systems. However, progress is being made today in the laboratory, in technology transition, in COTS products, and in standards. Although off-the-shelf middleware technology has not yet matured to cover the realm of large-scale, dynamically changing systems, DOC middleware has been applied to relatively small-scale and statically configured embedded systems [Sha98, NAS94]. Moreover, significant pioneering R&D on middleware patterns, frameworks, and standards for distributed systems has been conducted in the DARPA *Quorum* [Quorum99] and *PCES* [PCES02] programs, which played a leading role in:

1. Demonstrating the viability of integrating host infrastructure middleware, distribution middleware and common middleware services for DoD real-time embedded systems by providing foundation elements for managing key QoS attributes, such as real time behavior, dependability and system survivability, from a network-centric middleware perspective
2. Transitioning a number of new middleware perspectives and capabilities into DoD acquisition programs [Sha98, AegisOA, Holzer00] and commercially supported products and
3. Establishing the technical viability of collections of systems that can dynamically adapt within real-time constraints [Loy01] their collective behavior to varying operating conditions, in service of delivering the appropriate application level response under these different conditions.

3 FUTURE RESEARCH CHALLENGES AND STRATEGIES

In certain ways, each of the middleware successes mentioned in *Section 1 Introduction* can also be considered a partial failure, especially when viewed from a more complete perspective. In addition, other notable failures come from Air Traffic control, late opening of the Denver Airport, lack of integration of military systems causing misdirected targeting, and countless number of smaller, less visible systems which are cancelled, or are fielded but just do not work properly. More generally, connectivity among computers and between computers and physical devices, as well as connectivity options, is proliferating unabated, which leads to society's demand for network-centric systems of increasing scale and demanding precision to take advantage of the increased connectivity to better organize collective and group interactions/behaviors. Since these systems are growing (and will keep growing) their complexity is increasing, which motivates the need to keep application programming relatively independent of the complex issues of distribution and scale (in the form of advanced software engineering practices and middleware solutions). In addition, systems of national scale, such as the US air traffic control system or power grid, will of necessity be incremental and developed by many different organizations contributing to a common solution on an as yet undefined common high-level platform and engineering development paradigm.

Despite all the advances in the past decades, there are no mature engineering principles, solutions, or established conventions to enable large-scale, network-centric systems to be repeatably, predictably, and cost effectively created, developed, validated, operated, and enhanced. As a result, we are witnessing a complexity threshold that is stunting our ability to create large-scale, network-centric systems successfully. Some of the inherent properties that contribute to this complexity threshold include:

- Discrete platforms must be scaled to provide seamless end-to-end solutions
- Components are heterogeneous yet they need to be integrated seamlessly
- Most failures are only partial in that they effect subsets of the distributed components
- Operating environments and configurations are dynamically changing
- Large-scale systems must operate continuously, even during upgrades
- End-to-end properties must be satisfied in time and resource constrained environments
- Maintaining system-wide QoS concerns is expected

As described earlier, middleware resides between applications and the underlying OS, networks, and computing hardware. As such, one of its most immediate goals is to augment those interfaces with QoS attributes that serve as the linkage between application requirements and resource management strategies. Having a clear understanding of the QoS information is important so that it becomes possible to:

- Identify the users' (changeable) requirements at any particular point in time and
- Understand whether or not these requirements are being (or even can be) met.

This augmentation is beginning to occur, but largely on a component-by-component basis, not end-to-end. It is also essential to aggregate these requirements, making it possible to form decisions, policies, and mechanisms that begin to address a more global information management organization. Meeting these requirements will require new flexibility on the parts of both the application components and the resource management strategies used across heterogeneous systems of systems. A key direction for addressing these needs is through the concepts associated with managing adaptive behavior, recognizing that conditions are constantly changing and not all requirements can be met all of the time, yet still ensuring predictable and controllable end-to-end behavior.

Ironically, there is little or no scientific underpinning for QoS-enabled resource management, despite the demand for it in most distributed systems [Narain01]. Designers of today's complex distributed systems develop concrete plans for creating global, end-to-end functionality. These plans contain high-level abstractions and doctrine associated with resource management algorithms, relationships between these, and operations upon these. There are few techniques and tools that enable *users (e.g.,* commanders, administrators, and operators), *developers (e.g.,* systems engineers and application designers), and/or *applications,* to express such plans systematically, and to have these plans integrated and enforced automatically for managing resources at multiple levels in network-centric embedded systems.

Although there are no well accepted standards in these areas, work is progressing toward better understanding of these issues. To achieve these goals, middleware technologies and tools need to be based upon some type of layered architecture that is imbued with QoS adaptive middleware services. Figure 2 illustrates one such approach that is based on the Quality Objects (QuO) [ZBS97] project.

Middleware for Distributed Systems: State of the Art

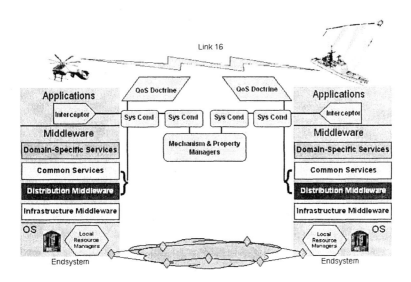

Figure 2. Decoupling Functional and QoS Attribute Paths

The QuO project and empirical demonstrations based on QuO middleware [Loy01, WMDS] are an example of how one might organize such a layered architecture designed to manage and package adaptive QoS capabilities [Sch02A] as common middleware services. The QuO architecture decouples DOC middleware and applications along the following two dimensions:

- *Functional paths,* which are flows of information between client and remote server applications. In distributed systems, middleware ensures that this information is exchanged efficiently, predictably, scalably, dependably, and securely between remote peers. The information itself is largely application-specific and determined by the functionality being provided (hence the term "functional path").
- *QoS attribute paths*, which are responsible for determining how well the functional interactions behave end-to-end with respect to key distributed system QoS properties, such as
    - How and when resources are committed to client/server interactions at multiple levels of distributed systems
    - The proper application and system behavior if available resources are less than the expected resources and
    - The failure detection and recovery strategies necessary to meet end-to-end dependability requirements.

In the architecture shown in Figure 2, the QuO middleware is responsible for collecting, organizing, and disseminating QoS-related meta-information that is needed to

1. Monitor and manage how well the functional interactions occur at multiple levels of distributed systems and
2. Enable the adaptive and reflective decision-making needed to support QoS attribute properties robustly in the face of rapidly changing mission requirements and environmental conditions.

In next-generation distributed systems, separating systemic QoS attribute properties from the functional application properties will enable the QoS properties and resources to change independently, *e.g.*, over different distributed system configurations for the same application, and despite local failures, transient overloads, and dynamic functional or QoS reconfigurations. An increasing number of next-generation applications will be developed as distributed "systems of systems," which include many interdependent levels, such as network/bus interconnects, local and remote endsystems, and multiple layers of common and domain-specific middleware. The desirable properties of these systems of systems include predictability, controllability, and adaptability of operating characteristics for applications with respect to such features as time, quantity of information, accuracy, confidence, and synchronization. All these issues become highly volatile in distributed systems of systems, due to the dynamic interplay of the many interconnected parts. These parts are often constructed in a similar way from smaller parts.

To address the many competing design forces and runtime QoS demands, a comprehensive methodology and environment is required to dependably compose large, complex, interoperable DOC applications from reusable components. Moreover, the components themselves must be sensitive to the environments in which they are packaged. Ultimately, what is desired is to take components that are built independently by different organizations at different times and assemble them to create a complete system. In the longer run, this complete system becomes a component embedded in still larger systems of systems. Given the complexity of this undertaking, various tools and techniques are needed to configure and reconfigure these systems, perhaps hierarchically, so they can adapt to a wider variety of situations.

The advent of open DOC middleware standards, such as CORBA and Java-based technologies, is hastening industry consolidation towards portable and interoperable sets of COTS products that are readily available for purchase or open-source acquisition. These products are still deficient and/or immature, however, in their ability to handle some of the important attributes needed to support future systems, especially mision critical and embedded distributed systems. Key attributes include end-to-end QoS, dynamic property tradeoffs, extreme scaling (large and small), highly mobile environments, and a variety of other inherent complexities. As the uses and environments for distributed systems grow in complexity, it may not be possible to sustain the composition and integration perspective we have achieved with current middleware platforms without continued R&D. Even worse, we may plunge ahead with an inadequate knowledge base, reverting to a myriad of high-risk independent solutions to common problems.

An essential part of what is needed to build the type of systems outlined above is the integration and extension of ideas that have been found traditionally in network management, data management, distributed operating systems, and object-oriented programming languages. We must create and deploy middleware-oriented solutions and engineering principles as part of the commonly available new, network-centric software infrastructure that is needed to develop many different types of large-scale systems successfully. The payoff will be reusable DOC middleware that significantly simplifies and reduces the inherent risks in building applications for complex systems of systems environments.

The remainder of this section presents an analysis of the challenges and opportunities for next-generation middleware and outlines the promising research strategies that can help to overcome the challenges and realize the opportunities.

3.1 Specific R&D Challenges

An essential part of what is needed to alleviate the inherent complexities outlined in the discussions above is the integration and extension of ideas that have been found traditionally in network management, data management, distributed operating systems, and object-oriented programming languages. The return on investment will yield reusable middleware that significantly simplifies the development and evolution of complex network-centric systems. The following are specific R&D challenges associated with achieving this payoff:

3.1.1 Providing end-to-end QoS support, not just component-level QoS

This area represents the next great wave of evolution for advanced middleware. There is now widespread recognition that effective development of large-scale network-centric applications requires the use of COTS infrastructure and service components. Moreover, the usability of the resulting products depends heavily on the properties of the whole as derived from its parts. This type of environment requires *predictable*, *flexible*, and *integrated* resource management strategies, both within and between the pieces, that are understandable to developers, visible to users, and certifiable to system owners. Despite the ease of connectivity provided by middleware, however, constructing integrated systems remains hard since it requires significant customization of non-functional QoS properties, such as predictable latency, dependability, and security. In their most useful forms, these properties extend end-to-end and thus have elements applicable to

- The network substrate
- The platform operating systems and system services
- The programming system in which they are developed
- The applications themselves and
- The middleware that integrates all these elements together.

The need for autonomous and time-critical behavior necessitates more flexible system infrastructure components that can adapt robustly to dynamic end-to-end changes in application requirements and environmental conditions. Next-generation applications will require the simultaneous satisfaction of multiple QoS properties, such as predictable latency/jitter/throughput, scalability, dependability, and security. Applications will also need different levels of QoS under different configurations, environmental conditions, and costs, and multiple QoS properties must be coordinated with and/or traded off against each other to achieve the intended application results. Improvements in current middleware QoS and better control over underlying hardware and software components–as well as additional middleware services to coordinate these–will all be needed.

Two basic premises underlying the push towards end-to-end QoS support mediated by middleware are that different levels of service are possible and desirable under different conditions and costs and the level of service in one property must be coordinated with and/or traded off against the level of service in another to achieve the intended overall results.

3.1.2 Adaptive and reflective solutions that handle both variability and control

It is important to avoid "all or nothing" point solutions. Systems today often work well as long as they receive all the resources for which they were designed in a timely fashion, but fail completely under the slightest anomaly. There is little flexibility in their behavior, i.e., most of the adaptation is pushed to end-users or administrators. Instead of hard failure or indefinite waiting, what is required is either *reconfiguration* to reacquire the needed resources automatically or *graceful degradation* if they are not available. Reconfiguration and operating under less than optimal conditions both have two points of focus: individual and aggregate behavior. Moreover, there is a need for interoperability of control and management mechanisms needed to carry out such reconfiguration. To date interoperability concerns have focused on data interoperability and invocation interoperability across components. Little work has focused on mechanisms for controlling the overall behavior of the end-to-end integrated systems. "Control interoperability" is needed to complement data and invocation interoperability if we are to achieve something more than a collection of independently operating components. There are requirements for interoperable control capabilities to appear in the individual resources first, after which approaches can be developed to aggregate these into acceptable global behavior through middleware based multi-platform aggregate resource management services.

To manage the broader range of QoS demands for next-generation network-centric applications, middleware must become more adaptive and reflective [ARMS01]. *Adaptive middleware* [Loy01] is software whose functional and QoS-related properties can be modified either:

- *Statically*, e.g., to reduce footprint, leverage capabilities that exist in specific platforms, enable functional subsetting, and minimize hardware/software infrastructure dependencies or
- *Dynamically*, e.g., to optimize system responses to changing environments or requirements, such as changing component interconnections, power levels, CPU/network bandwidth, latency/jitter; and dependability needs.

In mission-critical systems, adaptive middleware must make such modifications dependably, *i.e.*, while meeting stringent end-to-end QoS requirements. *Reflective middleware* [Bla99] goes further to permit automated examination of the capabilities it offers, and to permit automated adjustment to optimize those capabilities. Reflective techniques make the internal organization of systems–as well as the mechanisms used in their construction–both visible and manipulatible for middleware and application programs to inspect and modify at run-time. Thus, reflective middleware supports more advanced adaptive behavior and more dynamic strategies keyed to current circumstances, *i.e.*, necessary adaptations can be performed autonomously based on conditions within the system, in the system's environment, or in system QoS policies defined by end-users.

3.1.3 Combining model-integrated computing with DOC middleware

It has been increasingly recognized that source code is a poor way to document distributed system designs. Starting from informal design documentation techniques, such as flow-charts, model-integrated computing (MIC) is evolving DOC middleware towards more formal, semantically rich high-level design languages, and toward systematically capturing core aspects of designs via patterns, pattern languages, and architectural styles [MIC97]. MIC technologies are expanding their focus beyond application functionality to specify application quality of service (QoS) requirements, such as real-time deadlines and dependability constraints. These model-based tools provide application and middleware developers and integrators with higher levels of abstraction and productivity than traditional imperative programming languages provide. The following are some of the key R&D challenges associated with combining MIC and DOC middleware:

- Determine how to overcome problems with earlier-generation CASE environments that required the modeling tools to generate all the code. Instead, the goal should be to compose large portions of distributed applications from reusable, prevalidated DOC middleware components.
- Enhancing MIC tools to work in distributed environments where run-time procedures and rules change at rapid pace, e.g., by synthesizing, assembling, and validating newer extended components that conform to new rules that arise after a distributed system has been fielded.
- Devising domain-specific MIC languages that make DOC middleware more flexible and robust by automating the configuration of many QoS-critical aspects, such as concurrency, distribution, transactions, security, and dependability. This MIC-synthesized code may be needed to help bridge interoperability and portability problems between different middleware for which standard solutions do not yet exist.
- Train MIC tools to model the interfaces among various components in terms of standard middleware, rather than language-specific features or proprietary APIs.

3.1.4 Toward more universal use of standard middleware

Today, it is too often the case that a substantial percentage of the effort expended to develop applications goes into building *ad hoc* and proprietary middleware substitutes, or additions for missing middleware functionality. As a result, subsequent composition of these *ad hoc* capabilities is either infeasible or prohibitively expensive. One reason why redevelopment persists is that it is still often relatively easy to pull together a minimalist *ad hoc* solution, which remains largely invisible to all except the developers.

Unfortunately, this approach can yield substantial recurring downstream costs, particularly for complex and long-lived network-centric systems.

3.1.5 Leveraging and extending the installed base

In addition to the R&D challenges outlined above there are also pragmatic considerations, including incorporating the interfaces to various building blocks that are already in place for the networks, operating systems, security, and data management infrastructure, all of which continue to evolve independently. Ultimately, there are two different types of resources that must be considered:

1. Those that will be fabricated as part of application development and
2. Those that are provided and can be considered part of the substrate currently available.

While not much can be done in the short-term to change the direction of the hardware and software substrate that's installed today, a reasonable approach is to provide the needed services at higher levels of (middleware-based) abstraction. This architecture will enable new components to have properties that can be more easily included into the controllable applications and integrated with each other, leaving less lower-level complexity for application developers to address and thereby reducing system development and ownership costs. Consequently, the goal of next-generation middleware is not simply to build a better network or better security in isolation, but rather to pull these capabilities together and deliver them to applications in ways that enable them to realize this model of adaptive behavior with tradeoffs between the various QoS attributes. As the evolution of the underlying system components change to become more controllable, we can expect a refactoring of the implementations underlying the enforcement of adaptive control.

3.2 Fundamental Research Concepts

The following four concepts are central to addressing the R&D challenges described above:

3.2.1 Contracts and adaptive meta-programming

Information must be gathered for particular applications or application families regarding user requirements, resource requirements, and system conditions. Multiple system behaviors must be made available based on what is best under the various conditions. This information provides the basis for the contracts between users and the underlying system substrate.

These contracts provide not only the means to specify the degree of assurance of a certain level of service, but also provide a well-defined, high-level middleware abstraction to improve the visibility of adaptive changes in the mandated behavior.

3.2.2 Graceful degradation

Mechanisms must also be developed to monitor the system and enforce contracts, providing feedback loops so that application services can degrade gracefully (or augment) as conditions change, according to a prearranged contract governing that activity. The initial challenge here is to establish the idea in developers' and users' minds that multiple behaviors are both feasible and desirable. The next step is to put into place the additional middleware support–including connecting to lower level network and operating system enforcement mechanisms–necessary to provide the right behavior effectively and efficiently given current system conditions.

3.2.3 Prioritization and physical world constrained load invariant performance

Some systems are highly correlated with physical constraints and have little flexibility in some of their requirements for computing assets, including QoS. Deviation from requirements beyond a narrowly defined error tolerance can sometimes result in catastrophic failure of the system. The challenge is in meeting these *invariants* under varying load conditions. This often means guaranteeing access to some resources, while other resources may need to be diverted to insure proper operation. Generally collections of such components will need to be resource managed from a system (aggregate) perspective in addition to a component (individual) perspective.

3.2.4 Higher level design approaches, abstractions, and software development tools

Better techniques and more automated tools are needed to organize, integrate, and manage the software engineering paradigm and process used to construct the individual elements, the individual systems, and the systems of systems, without resorting to reimplementation for composition. Promising results in applying and embedding model-based software development practices [MIC97], as well as decomposition by aspects or views, suggest that applying similar design time approaches to QoS engineering may complement and make easier the runtime adaptation needed to control and validate these complex systems.

3.3 Promising Research Strategies

Although it is possible to satisfy contracts, achieve graceful degradation, and use modeling tools to globally manage some resources to a limited degree in a limited range of systems today, much R&D work remains. The research strategies needed to deliver these goals can be divided into the seven areas described below:

3.3.1 Individual QoS Requirements

Individual QoS deals with developing the mechanisms relating to the end-to-end QoS needs from the perspective of a single user or application. The specification requirements include multiple contracts, negotiation, and domain specificity. Multiple contracts are needed to handle requirements that change over time and to associate several contracts with a single perspective, each governing a portion of an activity. Different users running the same application may have different QoS requirements emphasizing different benefits and tradeoffs, often depending on current configuration. Even the same user running the same application at different times may have different QoS requirements, *e.g.*, depending on current mode of operation and other external factors. Such dynamic behavior must be taken into account and introduced seamlessly into next-generation distributed systems.

General negotiation capabilities that offer convenient mechanisms to enter into and control a negotiated behavior (as contrasted with the service being negotiated) need to be available as COTS middleware packages. The most effective way for such negotiation-based adaptation mechanisms to become an integral part of QoS is for them to be "user friendly," *e.g.*, requiring a user or administrator to simply provide a list of preferences. This is an area that is likely to become domain-specific and even user-specific. Other challenges that must be addressed as part of delivering QoS to individual applications include:

- Translation of requests for service among and between the various entities on the distributed end-to-end path
- Managing the definition and selection of appropriate application functionality and system resource tradeoffs within a "fuzzy" environment and
- Maintaining the appropriate behavior under composability.

Translation addresses the fact that complex network-centric systems are being built in layers. At various levels in a layered architecture the user-oriented QoS must be translated into requests for other resources at a lower level. The challenge is how to accomplish this translation from user requirements to system services. A logical place to begin is at the application/middleware boundary, which closely relates to the problem of matching application resources to appropriate distributed system resources. As system resources change in significant ways, either due to anomalies or load, tradeoffs between QoS attributes (such as timeliness, precision, and accuracy) may need to be (re)evaluated to ensure an effective level of QoS, given the circumstances. Mechanisms need to be developed to identify and perform these tradeoffs at the appropriate time. Last, but certainly not least, a theory of effectively composing systems from individual components in a way that maintains application-centric end-to-end properties needs to be developed, along with efficient implementable realizations of the theory.

3.3.2 Run-time Requirements

From a system lifecycle perspective, decisions for managing QoS are made at design time, at configuration/deployment time, and/or at run-time. Of these, the run-time requirements are the most challenging since they have the shortest time scales for decision-making, and collectively we have the least experience with developing appropriate solutions. They are also the area most closely related to advanced middleware concepts. This area of research addresses the need for run-time monitoring, feedback, and transition mechanisms to change application and system behavior, *e.g.,* through dynamic reconfiguration, orchestrating degraded behavior, or even off-line recompilation. The primary requirements here are *measurement*, *reporting*, *control*, *feedback*, and *stability*. Each of these plays a significant role in delivering end-to-end QoS, not only for an individual application, but also for an aggregate system. A key part of a run-time environment centers on a permanent and highly tunable measurement and resource status service as a common middleware service, oriented to various granularities for different time epochs and with abstractions and aggregations appropriate to its use for run-time adaptation.

In addition to providing the capabilities for enabling graceful degradation, these same underlying mechanisms also hold the promise to provide flexibility that supports a variety of possible behaviors, without changing the basic implementation structure of applications. This reflective flexibility diminishes the importance of many initial design decisions by offering late- and run-time-binding options to accommodate actual operating environments at the time of deployment, instead of only anticipated operating environments at design time. In addition, it anticipates changes in these bindings to accommodate new behavior.

3.3.3 Aggregate Requirements

This area of research deals with the system view of collecting necessary information over the set of resources across the system, and providing resource management mechanisms and policies that are aligned with the goals of the system as a whole. While middleware itself cannot manage system-level resources directly (except through interfaces provided by lower level resource management and enforcement mechanisms), it can provide the coordinating mechanisms and policies that drive the individual resource managers into domain-wide coherence. With regards to such resource management, policies need to be in place to guide the decision-making process and the mechanisms to carry out these policy decisions.

Areas of particular R&D interest include:

- *Reservations*, which allow resources to be reserved to assure certain levels of service
- *Admission control mechanisms*, which allow or reject certain users access to system resources
- *Enforcement mechanisms* with appropriate scale, granularity and performance and
- *Coordinated strategies and policies* to allocate distributed resources that optimize various properties.

Moreover, policy decisions need to be made to allow for varying levels of QoS, including whether each application receives guaranteed, best-effort, conditional, or statistical levels of service. Managing property composition is essential for delivering individual QoS for component based applications, and is of even greater concern in the aggregate case, particularly in the form of layered resource management within and across domains.

3.3.4 Integration Requirements

Integration requirements address the need to develop interfaces with key building blocks used for system construction, including the OS, network management, security, and data management. Many of these areas have partial QoS solutions underway from their individual perspectives. The problem today is that these partial results must be integrated into a common interface so that users and application developers can tap into each, identify which viewpoint will be dominant under which conditions, and support the tradeoff management across the boundaries to get the right mix of attributes. Currently, object-oriented tools working with DOC middleware provide end-to-end syntactic interoperation, and relatively seamless linkage across the networks and subsystems. There is no *managed* QoS, however, making these tools and middleware useful only for resource rich, best-effort environments.

To meet varying requirements for integrated behavior, advanced tools and mechanisms are needed that permit requests for *different* levels of attributes with different tradeoffs governing this interoperation. The system would then either provide the requested end-to-end QoS, reconfigure to provide it, or indicate the inability to deliver that level of service, perhaps offering to support an alternative QoS, or triggering application-level adaptation. For all of this to work together properly, multiple dimensions of the QoS requests must be understood within a common framework to translate and communicate those requests and services at each relevant interface. Advanced integration middleware provides this common framework to enable the right mix of underlying capabilities.

3.3.5 Adaptivity Requirements

Many of the advanced capabilities in next-generation information environments will require adaptive behavior to meet user expectations and smooth the imbalances between demands and changing environments. Adaptive behavior can be enabled through the appropriate organization and interoperation of the capabilities of the previous four areas. There are two fundamental types of adaptation required:

1. Changes beneath the applications to continue to meet the required service levels despite changes in resource availability and
2. Changes at the application level to either react to currently available levels of service or request new ones under changed circumstances.

In both instances, the system must determine if it needs to (or can) reallocate resources or change strategies to achieve the desired QoS. Applications need to be built in such a way that they can change their QoS demands as the conditions under which they operate change. Mechanisms for reconfiguration need to be put into place to implement new levels of QoS as required, mindful of both the individual and the aggregate points of view, and the conflicts that they may represent.

Part of the effort required to achieve these goals involves continuously gathering and instantaneously analyzing pertinent resource information collected as mentioned above. A complementary part is providing the algorithms and control mechanisms needed to deal with rapidly changing demands and resource availability profiles and configuring these mechanisms with varying service strategies and policies tuned for different environments. Ideally, such changes can be dynamic and flexible in handling a wide range of conditions, occur intelligently in an automated manner, and can handle complex issues arising from composition of adaptable components. Coordinating the tools and methodologies for these capabilities into an effective adaptive middleware should be a high R&D priority.

3.3.6 System Engineering Methodologies and Tools

Advanced middleware by itself will not deliver the capabilities envisioned for next-generation embedded environments. We must also advance the state of the system engineering discipline and tools that come with these advanced environments used to build complex distributed computing systems. This area of research specifically addresses the immediate need for system engineering approaches and tools to augment advanced middleware solutions. These include:

- *View-oriented or aspect-oriented programming techniques,* to support the isolation (for specialization and focus) and the composition (to mesh the isolates into a whole) of different projections or views of the properties the system must have. The ability to isolate, and subsequently integrate, the implementation of different, interacting features will be needed to support adapting to changing requirements.
- *Design time tools and model-integrated computing technologies,* to assist system developers in understanding their designs, in an effort to avoid costly changes after systems are already in place (this is partially obviated by the late binding for some QoS decisions referenced earlier).
- *Interactive tuning tools,* to overcome the challenges associated with the need for individual pieces of the system to work together in a seamless manner
- *Composability tools,* to analyze resulting QoS from combining two or more individual components
- *Modeling tools for developing system performance models* as adjunct means (both online and offline) to monitor and understand resource management, in order to reduce the costs associated with trial and error
- *Debugging tools,* to address inevitable problems.

3.3.7 Reliability, Trust, Validation, and Certifiability

The dynamically changing behaviors we envision for next-generation large-scale, network-centric systems are quite different from what we currently build, use, and have gained some degrees of confidence in. Considerable effort must therefore be focused on validating the correct functioning of the adaptive behavior, and on understanding the properties of large-scale systems that try to change their behavior according to their own assessment of current conditions, before they can be deployed. But even before that, longstanding issues of adequate reliability and trust factored into our methodologies and designs using off-the-shelf components have not reached full maturity and common usage, and must therefore continue to improve. The current strategies organized around anticipation of long life cycles with minimal change and exhaustive test case analysis are clearly inadequate for next-generation dynamic systems with stringent QoS requirements.

4 CONCLUDING REMARKS

In this age of IT ubiquity, economic upheaval, deregulation, and stiff global competition it has become essential to decrease the cycle-time, level of effort, and complexity associated with developing high-quality, flexible, and interoperable large-scale, network-centric systems. Increasingly, these types of systems are developed using reusable software (middleware) component services, rather than being implemented entirely from scratch for each use. Middleware was invented in an attempt to help simplify the software development of large-scale, network-centric computing systems, and bring those capabilities within the reach of many more developers than the few experts at the time who could master the complexities of these environments. Complex system integration requirements were not being met from either the *application perspective*, where it was too difficult and not reusable, or the *network or host operating system perspectives*, which were necessarily concerned with providing the communication and endsystem resource management layers, respectively.

Over the past decade, distributed object computing (DOC) middleware has emerged as a set of software protocol and service layers that help to solve the problems specifically associated with heterogeneity and interoperability. It has also contributed considerably to better environments for building network-centric applications and managing their distributed resources effectively. Consequently, one of the major trends driving researchers and practitioners involves

1. Moving toward a multi-layered architecture (i.e., applications, middleware, network and operating system infrastructure), which is oriented around application composition from reusable components, and
2. Moving away from the more traditional architecture, where applications were developed directly atop the network and operating system abstractions.

This middleware-centric, multi-layered architecture descends directly from the adoption of a network-centric viewpoint brought about by the emergence of the Internet and the componentization and commoditization of hardware and software.

Successes with early, primitive middleware has led to more ambitious efforts and expansion of the scope of these middleware-oriented activities, so we now see a number of distinct layers of the middleware itself taking shape. The result has been a deeper understanding of the large and growing issues and potential solutions in the space between complex distributed application requirements and the simpler infrastructure provided by bundling existing network systems, operating systems, and programming languages. Network-centric systems today are constructed as a series of layers of intertwined technical capabilities and innovations. The main emphasis at the lower middleware layers is in providing standardized core computing and communication resources and services that drive network-centric computing: overlays for the individual computers, the networks, and the operating systems that control the individual host and the message level communication.

At the upper layers, various types of middleware are starting to bridge the previously formidable gap between the lower-level resources and services and the abstractions that are needed to program, organize, and control systems composed of coordinated, rather than isolated, components. Key new capabilities in the upper layers include common and domain-specific middleware services that

- Enforce real-time behavior across computational nodes
- Manage redundancy across elements to support dependable computing and
- Provide coordinated and varying security services on a system wide basis, commensurate with the threat

There are significant limitations with regards to building these more complex systems today. For example, applications have increasingly more stringent QoS requirements. We are also discovering that more things need to be integrated over conditions that more closely resemble a volatile, changing Internet, than they do a stable backplane. Adaptive and reflective middleware systems [ARMS01] are a key emerging paradigm that will help to simplify the development, optimization, validation, and integration for distributed systems.

One problem is that the playing field is changing constantly, in terms of both resources and expectations. We no longer have the luxury of being able to design systems to perform highly specific functions and then expect them to have life cycles of 20 years with minimal change. In fact, we more routinely expect systems to behave differently under different conditions, and complain when they just as routinely do not. These changes have raised a number of issues, such as end-to-end oriented adaptive QoS, and construction of systems by composing off-the-shelf parts, many of which have promising solutions involving significant new middleware-based capabilities and services.

In the brief space of this paper, we can do little more than summarize and lend perspective to the many activities, past and present, that contribute to making DOC middleware technology an area of exciting current development, along with considerable opportunity and unsolved challenging problems. We have provided many references to other sources to obtain additional information about ongoing activities in this area. We have also provided a more detailed discussion and organization for a collection of activities that we believe represent the most promising future R&D directions of middleware for large-scale, network-centric systems. Downstream, the goals of these R&D activities are to:

1. Reliably and repeatably construct and compose network-centric systems that can meet and adapt to more diverse, changing requirements/environments and
2. Enable the affordable construction and composition of the large numbers of these systems that society will demand, each precisely tailored to specific domains.

To accomplish these goals, we must overcome not only the technical challenges, but also the educational and transitional challenges, and eventually master and simplify the immense complexity associated with these environments, as we integrate an ever growing number of hardware and software components together via DOC middleware and advanced network-centric infrastructures.

5 ACKNOWLEDGEMENTS

We would like to thank Don Hinton, Joe Loyall, Jeff Parsons, Andrew Sutton, Franklin Webber, and members of the Large-scale, Network-centric Systems working group at the Software Design and Producitivity workshop at Vanderbilt University, December 13-14, 2001 for comments that helped to improve this paper. Thanks also to members of the Cronus, ACE, TAO, and QuO user communities who have helped to shape our thinking on DOC middleware for over a decade.

6 REFERENCES

[AegisOA] Guidance Document for Aegis Open Architecture Baseline Specification Development, Version 2.0 (Draft), 5 July 2001.

[ARMS01] Schmidt D., Schantz R., Masters M., Sharp D., Cross J., and DiPalma L., "Towards Adaptive and Reflective Middleware for Network-Centric Combat Systems, Crosstalk, November 2001.

[Beck00] Beck K., *eXtreme Programming Explained: Embrace Change*, Addison-Wesley, Reading, MA, 2000.

[Ber96] Bernstein, P., "Middleware, A Model for Distributed System Service'", *Communications of the ACM*, 39:2, February 1996.

[Bla99] Blair, G.S., F. Costa, G. Coulson, H. Duran, et al, "The Design of a Resource-Aware Reflective Middleware Architecture", *Proceedings of the 2nd International Conference on Meta-Level Architectures and Reflection*, St.-Malo, France, Springer-Verlag, LNCS, Vol. 1616, 1999.

[Bol00] Bollella, G., Gosling, J. "The Real-Time Specification for Java," *Computer*, June 2000.

[Box97] Box D., *Essential COM*, Addison-Wesley, Reading, MA, 1997.

[Bus96] Buschmann, F., Meunier R., Rohnert H., Sommerlad P., Stal M., *Pattern-Oriented Software Architecture- A System of Patterns*, Wiley and Sons, 1996

[Chris98] Christensen C., *The Innovator's Dilemma: When New Technology Causes Great Firms to Fail*, 1997.

[Cuk98] Cukier, M., Ren J., Sabnis C., Henke D., Pistole J., Sanders W., Bakken B., Berman M., Karr D. Schantz R., "AQuA: An Adaptive Architecture that Provides Dependable Distributed Objects ", *Proceedings of the 17th IEEE Symposium on Reliable Distributed Systems*, pages 245-253, October 1998.

[Doe99] Doerr B., Venturella T., Jha R., Gill C., Schmidt D. "Adaptive Scheduling for Real-time, Embedded Information Systems," *Proceedings of the 18th IEEE/AIAA Digital Avionics Systems Conference (DASC)*, St. Louis, Missouri, October 1999.

[Gam95] Gamma E., Helm R., Johnson R., Vlissides J., *Design Patterns: Elements of Reusable Object-Oriented Software*, Addison-Wesley, 1995.

[Holzer00] Holzer R., "U.S. Navy Looking for More Adaptable Aegis Radar," *Defense News*, 18 September 2000.

[John97] Johnson R., "Frameworks = Patterns + Components", *Communications of the ACM*, Volume 40, Number 10, October, 1997.

[JVM97] Lindholm T., Yellin F., *The Java Virtual Machine Specification*, Addison-Wesley, Reading, MA, 1997.

[Kar01] Karr DA, Rodrigues C, Loyall JP, Schantz RE, Krishnamurthy Y, Pyarali I, Schmidt DC. "Application of the QuO Quality-of-Service Framework to a Distributed Video Application," *Proceedings of the International Symposium on Distributed Objects and Applications*, September 18-20, 2001, Rome, Italy.

[Loy01] Loyall JL, Gossett JM, Gill CD, Schantz RE, Zinky JA, Pal P, Shapiro R, Rodrigues C, Atighetchi M, Karr D. "Comparing and Contrasting Adaptive Middleware Support in Wide-Area and Embedded Distributed Object Applications". *Proceedings of the 21st IEEE International Conference on Distributed Computing Systems (ICDCS-21)*, April 16-19, 2001, Phoenix, Arizona.

[MIC97] Janos Sztipanovits and Gabor Karsai, "Model-Integrated Computing," IEEE Computer, Volume 30, Number 4, April 1997.

[Narain01] Narain S., Vaidyanathan R., Moyer S., Stephens W., Parameswaran K., and Shareef A., "Middle-ware For Building Adaptive Systems via Configuration," *ACM Optimization of Middleware and Distributed Systems (OM 2001) Workshop*, Snowbird, Utah, June, 2001.

[NAS94] New Attack Submarine Open System Imple-mentation, Specification and Guidance, August 1994.

[NET01] Thai T., Lam H., *.NET Framework Essentials*, O'Reilly, 2001.

[Omg98a] Object Management Group, "Fault Tolerance CORBA Using Entity Redundancy RFP", OMG Document orbos/98-04-01 edition, 1998.

[Omg98b] Object Management Group, "CORBAServices: Common Object Service Specification," OMG Technical Document formal/98-12-31.

[Omg99] Object Management Group, "CORBA Component Model Joint Revised Submission," OMG Document orbos/99-07-01.

[Omg00] Object Management Group, "The Common Object Request Broker: Architecture and Specification Revision 2.4, OMG Technical Document formal/00-11-07", October 2000.

[Omg00A] Object Management Group. "Minimum CORBA," OMG Document formal/00-10-59, October 2000.

[Omg00B] Object Management Group. "Real-Time CORBA," OMG Document formal/00-10-60, October 2000.

[Omg01] Object Management Group, "Dynamic Scheduling Real-Time CORBA 2.0 Joint Final Submission," OMG Document orbos/2001-04-01.

[PCES02] The Programmable Composition of Embedded Software (PCES) Project, DARPA Information Exploitation Office. http://www.darpa.mil/ito/research/pces/index.html

[Quo01] *Quality Objects Toolkit v3.0 User's Guide*, chapter 9, available as http://www.dist-systems.bbn.com/tech/QuO/release/latest/docs/usr/doc/quo-3.0/html/QuO30UsersGuide.htm

[Quorum99] DARPA, *The Quorum Program*, http://www.darpa.mil/ito/research/quorum/index.html, 1999.

[RUP99] Jacobson I., Booch G., and Rumbaugh J., *Unified Software Development Process*, Addison-Wesley, Reading, MA, 1999.

[Sch86] Schantz, R., Thomas R., Bono G., "The Architecture of the Cronus Distributed Operating System", *Proceedings of the 6th IEEE International Conference on Distributed Computing Systems (ICDCS-6)*, Cambridge, Massachusetts, May 1986.

[Sch98] Schantz, RE, "BBN and the Defense Advanced Research Projects Agency", Prepared as a Case Study for America's Basic Research: Prosperity Through Discovery, A Policy Statement by the Research and Policy Committee of the Committee for Economic Development (CED), June 1998 (also available as: http://www.dist-systems.bbn.com/papers/1998/CaseStudy).

[Sch02A] Schantz, R., Loyall, J., Atighetchi, M., Pal, P., "Packaging Quality of Service Control Behaviors for Reuse", ISORC 2002, *The 5th IEEE International Symposium on Object-oriented Real-time distributed Computing*, April 29 - May 1, 2002, Washington, DC. ...

[Sch98a] Schmidt D., Levine D., Mungee S. "The Design and Performance of the TAO Real-Time Object Request Broker", *Computer Communications Special Issue on Building Quality of Service into Distributed Systems,* 21(4), pp. 294—324, 1998.

[Sch00a] Schmidt D., Kuhns F., "An Overview of the Real-time CORBA Specification," *IEEE Computer Magazine*, June, 2000.

[Sch00b] Schmidt D., Stal M., Rohnert H., Buschmann F., *Pattern-Oriented Software Architecture: Patterns for Concurrent and Networked Objects*, Wiley and Sons, 2000.

[Sch02] Schmidt D., Huston S., *C++ Network Programming: Resolving Complexity with ACE and Patterns*, Addison-Wesley, Reading, MA, 2002.

[Sch03] Schmidt D., Huston S., *C++ Network Programming: Systematic Reuse with ACE and Frameworks*, Addison-Wesley, Reading, MA, 2003.

[Sha98] Sharp, David C., "Reducing Avionics Software Cost Through Component Based Product Line Development", *Software Technology Conference*, April 1998.

[SOAP01] Snell J., MacLeod K., *Programming Web Applications with SOAP*, O'Reilly, 2001.

[Ste99] Sterne, D.F., G.W. Tally, C.D. McDonell et al, "Scalable Access Control for Distributed Object Systems", *Proceedings of the 8th Usenix Security Symposium*, August,1999.

[Sun99] Sun Microsystems, "Jini Connection Technology", http://www.sun.com/jini/index.html, 1999.

[TPA97] Sabata B., Chatterjee S., Davis M., Sydir J., Lawrence T., "Taxonomy for QoS Specifications," *Proceedings of Workshop on Object-oriented Real-time Dependable Systems (WORDS 97)*, February 1997.

[Tho98] Thomas, Anne "Enterprise JavaBeans Technology", http://java.sun.com/products/ejb/white_paper.html, Dec. 1998

[Wol96] Wollrath A., Riggs R., Waldo J. "A Distributed Object Model for the Java System," *USENIX Computing Systems*, 9(4), 1996.

Internet Routing—The State of the Art

K.Ramasamy, A.Arokiasamy, and P.A.Balakrishnan
Faculty Of Engineering, Multimedia University, Malaysia
E-mail: k.ramasamy@mmu.edu.my

Abstract: The Internet protocols are the world's most popular open-system (nonproprietary) protocol suite because they can be used to communicate across any set of interconnected networks. In this paper an overview on some internet routing is analyzed. The routing protocols used in the internet are ARPANET protocol, Routing Information Protocol (RIP), Open Shortest Path First protocol (OSPF), Interior Routing Protocol (IRP), Exterior Routing Protocol (ERP), Interior Gateway Protocol (IGP), Exterior Gateway Protocol (EGP), Transfer Control Protocol (TCP) and Internet Protocol (IP). The routing protocols are used based on reliability, scalability and security. The OSPF protocol has been given the status of being the standard routing protocol for IP internetworks. The routing protocol, which is currently in use, is Transmission Control Protocol/Internet Protocol (TCP/IP), which is providing to be a great success for transmission of data. TCP provides a reliable connection for the transfer of data between applications. The Internet Protocol (IP) is a network-layer protocol that contains addressing information and source control information that enables packets to be routed. It has two primary responsibilities: providing connectionless, best effort delivery datagrams through an internetwork and providing fragmentation and ressembly of datagrams to support data links with different maximum-transmission unit (MTU) sizes. In addition to the TCP, there is another transport level protocol, the UserDatagram Protocol (UDP) which provides connectionless service for applicationlevel procedures. UDP is useful when TCP would be too complex, too slow or just unnecessary. UDP address is the combination of 32-bit IP address and 16-bit port number. Unlike IP,it does checksum its data,ensuring data integrity.IPv6 is a new version of the Internet protocol based on IPv4.It increases the IP address size from 32 bits to 128 bits to support more levels of addressing hierarchy, much greater number of addressable modes and simpler auto-configuration of address.Scalability of multicast address is introduced.A new type of address called an anycast address is also defined to send a packet to any one of a group of nodes.

1 INTRODUCTION

Internet is a collection of communication networks interconnected by Bridges and/or Routers. Routing is the process of finding a path from a source to every destination in the network. The heart of communication depends on routing. One such communication is internet. Here we analyze about the routing which takes place in the internet. The process of Routing in an Internetwork is referred to as Internet Routing. The principal criterion of successful routing are, correctness, computational simplicity, stability, fairness, optimality, the most reliable route and the least expensive route. In general forwarding a packet from source to destination using the best path is known as optimal routing.

A Sub network refers to a constituent network of an Internet. A device attached to one of the sub networks of an internet that is used to support end-user applications or services is called as End Systems. A device used to connect two sub networks and permit communication between end systems attached to different sub networks is known as Intermediate Systems. Two important terminologies related to Internet routing is Bridge and Router. A Bridge operates at the Data-Link layer. A Bridge is an IS used to connect two LANs that use similar LAN protocols. The bridge acts as an address filter, picking up packets from one LAN that are intended for a destination on another destination on another LAN and passing those packets on. The bridge does not modify the contents of the packets and does not add anything to the packet.

A router is a special purpose computer that transfers data to and from the next higher level of the hierarchy. Or in other words a Router is defined as a host that has an interface on more than one network. The router mainly operates on the Network layer. The key to effective performance management in a packet-based inter network environment is the router. The router must

1. Have the processing capacity to move IP data grams through the router at extremely high rates.
2. Have sufficient knowledge of the networked configuration to pick a route that is appropriate to a given class of traffic and to compensate congestion and failure in the internet work.
3. Employ a scheme for exchanging routing information with other routers that is effective and does not excessively contribute to the traffic burden.
4. Packets from one network to another network may have to be broken in to smaller pieces, this is known as segmentation. For example, Ethernet imposes a maximum packet size of 1500 bytes, and a minimum packet size of 1000 bytes is common in x.25 networks. So, a router for

retransmission on a specific network may have to be fragmented in to smaller ones.
5. The hardware and software interfaces to various networks differ. The router must be independent of these differences.
6. The operation of the routers should not depend on an assumption of network reliability.

The ability to route traffic between the ever increasing number of networks that comprise the internet, has become a major problem. The routers in an internet perform much the same function as a packet-switching nodes (PSNS) in a packet-switching network. Just as the PSN is responsible for receiving and forwarding packets through a packet-switching network, the router is responsible for receiving and forwarding IP datagrams through an internet. The routers of an internet need to make routing decisions based on knowledge of the topology and conditions of the internet.

2 STATE OF THE ART

The problem of Routing in the internetworks has been studied for well over seven years. The subject has improved where a number of books have been written thoroughly examining different issues and solutions. A key distinction we will make concerning the study of routing is that between routing protocols, by which we mean mechanisms for disseminating routing information within a network and the particulars of how to use that information to forward traffic, and routing behavior, meaning how in practice the routing algorithms perform. Routing algorithms referred to as the network layer protocol that guides packets through the communication subnet to their correct destination.

Previously McQuillan has discussed about the initial ARPANET routing algorithm[1] and the algorithms that replaced it [2]; the Exterior Gateway Protocol used in the NSFNET [3] and the Border Gateway Protocol that replaced it[4]; the related work by Estrin et al., on routing between administrative domains [5]; Perlman and Verghese's discussion of difficulties in designing routing algorithms [6]; Perlman's comparison of the popular OSPF and IS-IS protocols [7]; Barasel et al's survey of routing techniques for very high speed networks [8].

Other Current works include End to End Routing behavior in the internet by Vern Paxson [9]; Chinoy's work on Dynamics on Internet routing information [10]; Deering and Cheriton's work on Multicast Routing in Datagram Internetworks and Extended LANs [11]; Stevens and Gomer's work on TCP/IP Protocols [12].

2.1 Current Trends

For Routing purpose, the Internet is partitioned in to a disjoint set of Autonomous systems (AS's), a notion first introduced in [3]. Originally an AS was a collection of routers and hosts unified by running a single "interior gateway protocol". Routing between autonomous systems provides the highest level of Internet interconnection. There are many techniques introduced after the ARPANET routing protocol. They are Routing information protocol(RIP), Open shortest path first protocol(OSPF), Interior routing protocol(IRP), Exterior routing protocol(ERP), Interior gateway protocol (IGP), Exterior gateway protocol(EGP), Transfer control protocol (TCP) and Internet protocol(IP).

All the above mentioned routing protocols are used for the purpose of intemet routing. Currently Internet protocol(IP) is used more for intemet routing on the basis of reliability, Scalability and security. We will discuss more about this in the later part.

3 ROUTING METHODS

There are different routing methods in existence [14]. Basic routing methods are described in the following sections.

3.1 Source Routing

All the information about how to get from here to there is first collected at the source, which puts it in to packets that it launches toward the destination. The job of the intervening network is simply to read the routing information from the packets and act on it faithfully.

3.2 Hop-by-Hop Routing

The source is not expected to have all the information about how to get from here to there ; it is sufficient for the source to know only how to get to the next hop, and for that system to know how to get to the next hop and so on until the destination is reached. The term "hop" is used to describe a unit of distance in routing algorithms.

3.3 Direct Routing

To see if the destination is on the same physical network:

a) Extract the NETID portion of the destination IP address.

b) Compare with source's NETID(S)

This method is extremely efficient because of the address structure. To send from node A to node B:

a) Encapsulate the datagram in a physical frame
b) Bind the IP address to the physical address (ARP, RARP)
c) Transmit the frame.

The final Gateway along the total route will always perform direct routing to the destination node.

3.4 Indirect Routing

- More difficult because it must be sent to an appropriate gateway (or default gateway).
- Gateways form a cooperative, interconnected structure which pass datagrams to each other until a gateway can deliver the datagram directly.

Indirect routing involves:

a) Software extraction of the datagrain from the frame.
b) Selecting an appropriate route based on routing algorithm used.
c) Re-encapsulation of datagram in to a new frame.
d) Setting the appropriate physical addresses in the frame header.
e) Transmit the new frame on to the network.

3.5 Static Routing

Static Routing is a type of routing that do not base their routing decisions on measurements or estimates of the current traffic and topology. The principle of static routing is used widely when there is an instance that routing decisions can be made independently.

3.6 Dynamic Routing

Dynamic routing, on the other hand, attempt to change their routing decisions to reflect changes in topology and the current traffic. The principle of dynamic routing is also used widely, because of its nature and tendency to change with the routing environment.

4   ROUTING PROTOCOL REVIEW

Routing Protocols are an essential ingredient to the operation of an internet. Internets operate on the basis of routers that forward IP datagrams from router to router on a path from the source host to the destination host. For a router to perform its function, it must have some idea of the topology of the internet and the best route to follow. It is the purpose of the Routing protocol to provide the needed information.

Routing is accomplished by means of routing protocols that establish consistent routing tables in every router in the network. Routing protocols are the base for Routing algorithms. For a routing protocol an algorithm is used called as routing algorithm.

Routing Protocols are chosen for the following criteria:

- Availability
- Speed of adaptation to finding alternate routes
- Speed in resolving routing loops
- Minimizing routing table space
- Minimizing control messages
- Overhead load placed on the network, by the protocol
- Scalability
- Security
- Robustness

4.1   Autonomous Systems

Routing protocols are solely based on Autonomous Systems. Generally End Systems are considered as Autonomous Systems which provides the base for routing protocols. The Autonomous System (AS) exhibits the following characteristics:

1. An AS consists of a group of routers exchanging information through a common routing protocol.
2. An AS is a set of routers and networks managed by a single organization.
3. Except in times of failure, an AS is connected (in a graph-theoretic sense).

There are two types of Routing Protocols in an Autonomous System: interior routing protocols and exterior routing protocols.

4.2 Interior Routing Protocol (IRP)

The protocol which is used to pass routing information between routers within an AS is referred to as Interior routing protocols.

Most important examples of IRP is

- Routing Information Protocol (RIP)
- Open Shortest Path First Protocol (OSPF)

4.3 Exterior Routing Protocol (ERP)

The Protocol which is used to pass routing information between routers in different ASs is referred to as an Exterior Routing Protocol. The application of exterior and interior routing protocols is shown in figure 1.

Examples of ERP is

- Border Gateway Protocol (BGP)
- Inter Domain Routing Protocol (IDRP)

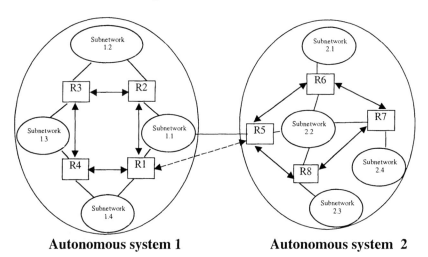

Figure 1: Application of exterior and interior routing protocols

4.4 Routing Information Protocol (RIP)

The Routing information protocol is known as distance vector protocol, which means that all of router's decisions about which path to use are based on distance The Bellman-ford algorithm is used by RIP for computing the shortest path. An example of RIP has been given. It is shown in figure 2.

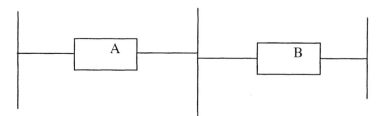

Fig.2. Basic Routing Example

Here Router A needs to compute a path to a destination that it is not directly connected to. Router A would like to get to network 3, which is on the other side of Router B. In this case where there are multiple paths to the same destination then each path will be examined and the path with the lowest cost, or number of hops, will be chosen as the path to use for that destination.

To Summarize, the cost of a path to get from A to F could be written as follows:

$$d(A,B) + d(B,C) + d(C,D) + d(D,E) + d(E,F)$$

Where $d(x,y)$ is the distance, or number of hops, from x to y. In most implementations of RIP each intermediate hop is assigned a cost or metric of 1, and if this is the case then the cost of getting from A to F is simply equal to the number of intermediate networks between the source and the destination. This formula is used in all cases of Distance Vector Routing. Generally Bellman-Ford algorithm is used for this kind of protocol. In some cases Ford-Fulkerson algorithm is also used.

A basic routing scenario is shown in figure 3.

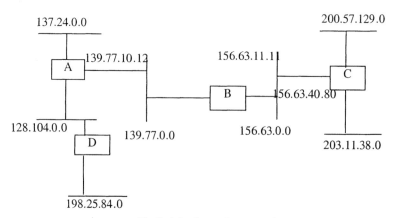

Fig.3. A basic routing scenario

4.4.1 Building a routing table

In the above example we have four routers and seven networks. The initial RIP routing table is shown in table.1.

Table 1.Initial RIP routing table

Destination	Next Hop	Metric	D/R	L/R	Interface
139.77.0.0	0.0.0.0	I	D	L	1
156.63.0.0	0.0.0.0	I	D	L	2

Here, Destination is the Destination network number. Next hop lists the complete IP address of the next hop router that is between you and the destination network. Metric is the hop count field, it lists how far away the particular destination is from you.

D/R: D   Directly connected network; R = Remote network.
L/R: L   Network was learned because it is local.

Interface is the logical interface number on which the routing information was received.

The completed RIP Routing Table is shown in table 2.

Table 2. Completed RIP Routing Table

Destination	Next Hop	Metric	Interface
128.104.0.0	139.77.10.12	2	1
137.24.0.0	139.77.10.12	2	1
139.77.0.0	0.0.0.0	1	1
156.63.0.0	0.0.0.0	1	2
198.25.84.0	139.77.10.12	3	1
200.57.129.0	156.63.40.80	2	2
203.11.38.0	156.63.40.80	2	2

4.4.2 RIP Message Format

The RIP message format is shown in figure 4.

Command (8)
Version (8)
Must be zero (16)
Address Family Identifier (16)
Must be zero (16)
IP Address (32)
Must be zero (64)
Metric (32)

Fig.4. RIP message format

The Description of RIP message format is given as follows

Command:	Defines whether this RIP message is a request (1) or a response (2).
Version:	The version of RIP being used. Current version of RIP being used is 1.
Address Family Identifier:	Identifier of the protocol that is using RIP for its routing Table calculations. The Address family identifier for IP is 2.
IP Address :	This is the IP address of the network that is being advertised or requested. This value may be just a network, or it may be full IP address.
Metric:	The number of hops to the destination network. This value will be between 1 and 15, with the value 16 used to set the metric to infinity.

4.4.3 RIP Limitations

1. Limited Network Diameter: This refers to the limited hop count allowed by RIP. Without the use of extended RIP (ERIP), your network design is limited to a maximum of fifteen routers between any two networks.
2. Convergence: The time it takes to learn about changes in the network can cause a number of problems because routers believe erroneous information longer than they should.
3. Aging of Entries: Entries in a routing table are allowed to remain in the routing table for 180 seconds in the absence of an update about the network.
4. Subnetting: It is important to remember that RIP assumes that the same subnet mask is being used throughout the entire RIP environment.

4.5 Open Shortest Path first Protocol (OSPF)

The Open shortest path first protocol has been given the status of being the standard routing protocol for IP internetworks. OSPF was created specifically for the IP environment to solve many problems inherent with RIP. Here the Algorithm used is Dijkstra's algorithm for finding the shortest path. A "path" is a series of steps needed to get from one point to another point in the internetwork.

OSPF uses the concept of areas and the backbone to allow us to divide the domain in to manageable entities. The procedure of getting from one area to another can be broken up in to three steps:

1. The path from the source network to the edge of the originating area.

2. The path from the originating area, across the backbone, to the edge of the destination area.
3. The path from the edge of the destination area to the destination network.

4.5.1 Shortest Path Tree

The Shortest path tree shows the best path to each destination in the internetwork with the associated total path cost. It is from the shortest path tree that we build a routing table. The shortest path tree for router11 is given as example, which is as shown in figure 5.

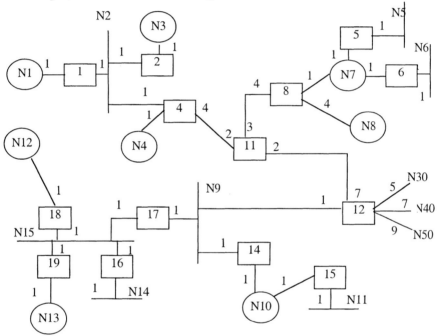

Fig.5. Shortest path tree for router11

A specific example has been given to show the shortest path area. The corresponding basic routing table is shown in table.3.

Dijkstra's algorithm is a graph theory algorithm that has been applied to the routing world, specifically to OSPF. A sample Dijkstra's algorithm graph is shown in figure 6.

Table 3. Corresponding basic routing table

Destination	Next Hop	Cost
NI	Router 4	4
N2	Router 4	3
N3	Router 4	4
N4	Router 4	3
N5	Router 8	5
N6	Router 8	5
N7	Router 8	4
N8	Router 8	4
N9	Router 12	3
N10	Router 12	4
Nll	Router 12	5
N12	Router 12	5
N13	Router 12	5
N14	Router 12	5
N15	Router 12	4
N30	Router 12	5
N40	Router 12	7
N50	Router 12	9

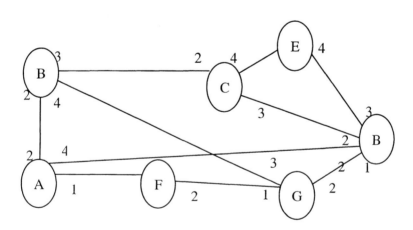

Fig.6. A sample Dijkstra's algorithm graph

The cost of a path between the source(S) and the destination(T) is:

d(S,N 1) + d(N 1,N2) + d(N2,N3) ... + d(Nx,T)

Where d(x,y) is the cost to get from point x to point y on the graph. The calculation is done based on the given formula.

4.5.2 Router Classifications:

The Router classifications for the OSPF is given with the illustrations. The OSPF defines four types of routers, they are as follows;

Internal Routers:	An Internal router is a router that has all of its interfaces connected to networks within the same area.
Area Border Routers:	An area border router is a router that has network connections to more than one area.
Backbone Routers:	A backbone router is a router that has interfaces connected to the backbone.
Autonomous System Boundary Routers:	Autonomous System Boundary routers are routers that have interfaces attached to network outside autonomous system.

The RIP is generally called as Distance vector Routing and the OSPF is generally called as Link state Routing. A comparison is made between the two important routing algorithms. It is shown in table.4.

Table.4. Comparison is made between DVR and LSR

Distance-Vector Routing	Link-State Routing
Each router sends routing information to its neighbors	Each router sends routing information to all other routers
The information sent is an estimate of its Path cost to all networks	The information sent is the exact value of its link cost to adjacent networks
Information is sent on a regular periodic Basis	Information is sent when changes occur
A router determines next-hop information by using the distributed Bellman-Ford algorithm on the received estimated path costs	A router first builds up a description of the topology of the internet and then may use any routing algorithm to determine next-hop information

5 LATEST ROUTING PROTOCOL

The Routing protocol which is currently in use is Transmission Control Protocol/ Internet Protocol (TCP/IP) [13]. This is proving to be a great success for transmission of data. TCP/IP is a result of protocol research and development conducted on the experimental packet-switched network, ARPANET, funded by the Defense Advanced Research Projects Agency (DARPA), and is generally referred to as the TCP/IP protocol suite. There is no official protocol model for TCP/IP like OSI. But based on the protocol standards, we can organize the communication task for TCP/IP into five relatively independent layers:

- Application layer
- Host-to-host, or Transport layer
- Internet layer
- Network access layer
- Physical layer.

5.1 TCP and UDP

For most applications running as part of the TCP/IP protocol architecture, the transmission layer protocol is TCP. TCP provides a reliable connection for the transfer of data between applications [13]. The figure.7. shows the header format for TCP;

Bit: 0	4	10	16		
Source Port			Destination Port		
Sequence number					20 Octets
Acknowledge number					
Header Length	Unused		Flags	Window	
Check sum				Urgent pointer	
Options + Padding					

TCP Header

Fig.7. Header format for TCP

The header format for TCP, which is a maximum of 20 octets, or 160 bits. The source port and destination port fields identify the applications at the source and the destination systems that are using this connection. The sequence number, Acknowledgement Number, and window fields provide flow control and error control.

In addition to the TCP, there is another transport level protocol : the user datagram protocol (UDP). UDP provides a connectionless service for applicationlevel procedures. The UDP header is shown in figure 8.

Bits: 0	16	32	
Source Port	Destination Port		8 Octets
Segment length	Check sum		

Fig.8. UDP Header

UDP enables a procedure to send messages to other procedures with a minimum of protocol mechanism.

5.1.1 Ipv4 and Ipv6

The keystone of the TCP/IP protocol architecture has been the IP [13] The figure 9 shows the lpv4 header format.

Bits: 0	4	8	16	19	31
Version	IHL	Type of service	Total Length		
Identification			Flags	Fragment Offset	
Time to live		Protocol	Header Checksum		
Source address					
Destination address					
Options + Padding					

Fig.9. Ipv4 header format

The above IP header format, which is a minimum of 20 octets, or 160 bits. The header includes 32-bit source and destination addresses. The protocol field indicates whether TCP, UDP, or some other higher-layer protocol is using IP. The figure 10. shows the header format of the latest lpv6 protocol.

The latest Ipv6 contains 40 octets, or 320 bits. There are lot of modifications has been done to improvise the latest version. lpv6 includes 128-bit source and destination address fields.

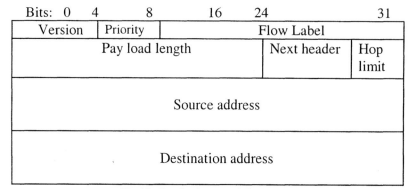

Fig.10. Header format of the latest Ipv6 protocol

5.1.2 Operation of TCP/IP

Figure 11 shows the concepts of TCP/IP [13]. It also indicates how these protocols are configured for communications.

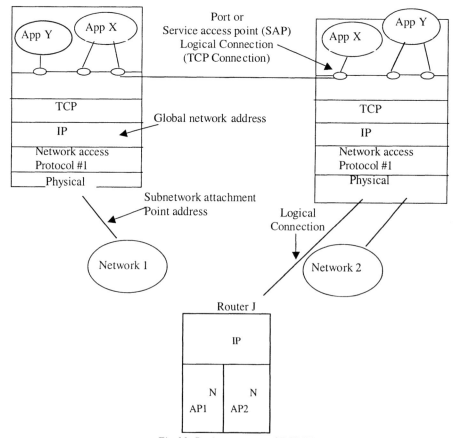

Fig.11. Basic concepts of TCP/IP

This Protocol enables the host to send data across the subnetwork to another host, in the case of a host on another subnetwork, to a router. IP is implemented in all end systems and routers. It acts as a relay to move a block of data from one host, through one or more routers, to another host. TCP is implemented only in the end systems; it keeps track of the blocks of data to assure that all are delivered reliably to the appropriate application. For successful communication, every entity in the overall system must have a unique address.

The protocol data units in the TCP/IP architecture is shown in figure 12.

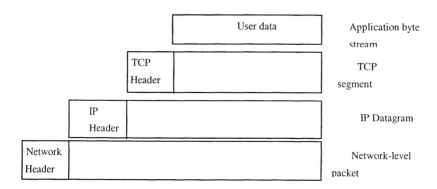

Fig.12. Protocol data units in the TCP/IP architecture

Here there are four important levels. The base most level is associated with the Network layer. The next level is associated with the Internet layer. The third level is associated with the Transport layer. The final level is associated with the Application layer. The TCP/IP architecture entirely depends on these four levels. The data communication takes place only with the cooperation of these four levels. To control the operation of TCP/IP, control information as well as user data must be transmitted. Let us assume that the sending process generates a block of data and passes this to TCP. TCP may break this block into smaller pieces to make it manageable. To each of these pieces, TCP appends control information in the TCP header, forming a TCP segment. The control information is to be used by the TCP protocol entity at host B.

5.2 Comparison with Previous Protocol

After the analysis of TCP/IP we can clearly see the advantages over OSI. A comparison of the TCP/IP and OSI protocol is shown in figure 13.

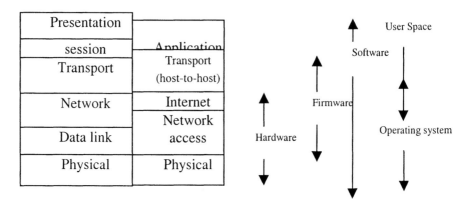

Fig.13. Comparison of the TCP/IP and OSI protocol

The Overall Operation of TCP/IP, that is its Action at Sender, Action at Router and its Action at Receiver is shown in a brief way to see the exact process.

The figure 14 shows about Action at Sender[13];

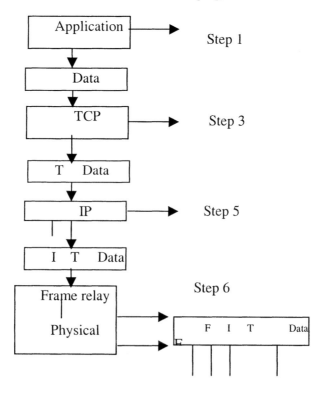

Fig.14.The Action at Sender

The operation of TCP/IP action at sender is described in the following steps with peer-to-peer dialogue;

Step 1: Preparing the data: The application protocol prepares a block of data for transmission. For example, an e-mail message (SMTP), a file (FTP), or a block of user input (TELNET).

Peer-to-peer dialogue: Before data are sent, the sending and receiving applications agree on format and encoding and agree to exchange data.

Step 2: Using a common syntax : If necessary, the data are converted to a form expected by the destination. This may include a different character code, the use of encryption, and/or compression.

Step 3: Segmenting the data: TCP may break the data block into a number of segments, keeping track of their sequence. Each TCP segment includes a header containing a sequence number and a frame check sequence to detect errors. Peer-to-peer dialogue: The two TCP entities agree to open a connection.

Step 4: Duplicating segments: A copy is made of each TCP segment, in case the loss or damage of a segment necessities retransmission. When an acknowledgement is received from the other TCP entity, a segment is erased.

Step 5: Fragmenting the segments: IP may break a TCP segment into a number of datagrams to meet size requirements of the intervening networks. Each datagram includes a header containing a destination address, a frame check sequence, and other control information.

Peer-to-peer dialogue: Each IP datagram is forwarded through networks and routers to the destination system.

Step 6: Framing: A frame relay header and trailer is added to each IP datagram. The header contains a connection identifier and the trailer contains a frame check sequence.

Peer-to-peer dialogue: Each frame is forwarded through the frame relay network.

Step 7: Transmission: Each frame is transmitted over the medium as a sequence of bits.The Action at router[13] is clearly analyzed in the following figure.15.

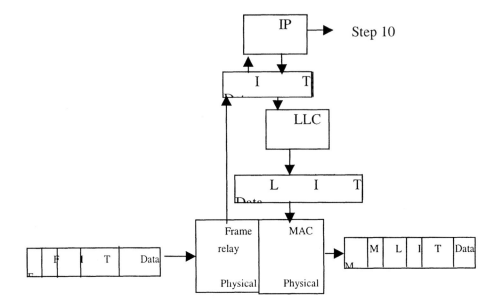

Fig.15. The Action at router

The operation of TCP/IP Action at router is described in the following steps;

Step 8: Arriving at router: The incoming signal is received over the transmission medium and interpreted as a frame of bits.

Step 9: Processing the frame: The frame relay removes the header and trailer and processes them. The frame check sequence is used for error detection. The connection number identifies the source.

Step 10: Routing the Packet: IP examines the IP header and makes a routing decision. It determines which outgoing link is to be used and then passes the datagram back to the link layer for transmission on that link.
Peer-to-peer dialogue: The router will pass this datagram onto another router or to the destination system.

Step 11: Forming LLC PDU: An LLC header is added to each IP datagram to form an LLC PDU. The header contains sequence number and address information.

Step 12: Framing: A MAC header and trailer is added to each LLC PDU, forming a MAC frame. The header contains address information and the trailer contains a frame check sequence.

Step 13: Transmission: Each frame is transmitted over the medium as a sequence of bits.

The Action at Receiver [13] is shown in figure 16.

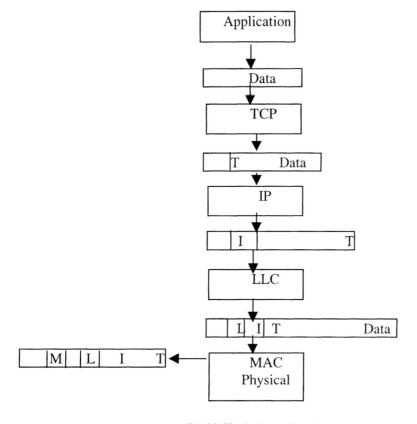

Fig.16. The Action at Receiver

The operation of TCP/IP action at receiver is described in the following steps;

Step 14: Arriving at destination: The incoming signal is received over the transmission medium and interpreted as a frame of bits.

Step 15: Processing the frame: The MAC layer removes the header and trailer and processes them. The frame check sequence is used for error detection.

Step 16: Processing the frame: The LLC layer removes the header and processes it. The sequence number is used for flow and error control.

Step 17: Processing the IP datagram: IP removes the header. The frame check sequence and the other control information are processed.

Step 18: Processing the TCP segment: TCP removes the header. It checks the frame check sequence and acknowledges if there is a match and discards for mismatch. Flow control is also performed.

Step 19: Reassembling user data: If TCP has broken the user data into multiple segments, these are reassembled and the block is passed up to the application.

Step 20: Delivering the data: The application performs the needed transformations, including decompression and decryption, and directs the data to the appropriate file or other destination.

6 OVERVIEW OF ROUTING PROTOCOLS

The overall comparison of some Routing Protocols and its characteristics are shown in the following table.5.

7 CONCLUSION

The Routing in the Internet is clearly one of the important problems faced during this era. There are some few important routing algorithms that essentially fit for internet routing. Still there are some more algorithms which can ease the problems of intemet routing. Global information, by its very nature, is hard to collect, subject to frequent change, and voluminous. How can we summarize this information to extract only the portions relevant to each node, this lies in the heart of Routing Protocols.

We conclude that any routing protocol must communicate global topological information to each routing element to allow it to make local routing decisions. Global Optimization is very much possible through Internet Routing. This is possible only based on the routing protocols.

8 FUTURE WORK

There are some Efficient Loop free routing algorithm reported to date and perhaps to find the shortest path. A problem that needs to be investigated is how to ensure that routers using shortest path will never update their routing information using stale distance vectors. Other interests could be to simulate data traffic and measure the resulting packet loss from looping.

Table.5. Overall comparison of some Routing Protocols and its characteristics

Character-istics	Routing Protocols							
	RIP	IGRP	OSPF	ISO IS-IS ISO ES-IS	EGP	BGP	RTMP	DEC net
Static Vs. Dynamic	D	D	D	D	D	D	D	D
Distributed Vs. Centralized	D	D	D	D	D	D	D	D
Single Vs. Multi-path	S	M	M	ES-IS= SS IS-IS=M	M	S	M	
Flat Vs. Hierarchical	F	F	H	ES-IS= F IS-IS=H	F	F	F	H
Source Vs. Router Intelligent	R	R	R	R	R	R	R	R
Intra- Vs. Inter-domain	Intra	Intra	Intra	ES-IS= Intra IS-IS= both	Inter	Inter	Inter	Both
Link state Vs. Distance Vector	DV	DV	IS-IS =LS	ES –IS = Neither	Neither (Reach-ability)	Neither (Reach-ability)	DV	LS
Yrs. Since Intial Specification	10	6	2	5	9	2	7	12
Metric Factors considered	Hop-count Delay Bandwidth Load MTU Range: 1-16	Reliability Unit Range: 1-255. Number Optional TOS Service Cost	Arbitrary IS-IS=Arbitrary Unit Number Range: 1-65536	ES-IS = ---- Range: 1-1024	Unit Number	Arbi-trary	Hop-count Unit Number Range: 1-16	Arbi-trary Range 1-1022

Because of the growing size of the internetworks, it would be of great interest to extend shortest path algorithms to hierarchical networks. A promising approach to address this problem is to adopt McQuillan's scheme to hierarchical routing. Some simulation results of hierarchical routing scheme should outperform OSPF.

Research on Internet Routing to transmit Audio and Video is well in progress. In future the study of routing in streams should be done by minimizing the delays in the frame relays.

9 REFERENCES

[1] J.McQuillan, G.Falk and I.Ritcher, " A Review of the development and Performance of the ARPANET Routing Algorithm", IEEE Transactions on Communications, 26(12), pp.1802-181 1, December 1978.
[2] J.McQuillan, I.Ritcher and E.Rosen, "The New routing algorithm for the ARPANET", IEEE Transactions on communcations, 28(5), pp.711-719, May 1980.
[3] E.Rosen, "Exterior Gateway Protocol (EGP)", RFC 896, Network Information Center, SRI International, Menlo Park, CA, October 1982.
[4] Y. Rekhter, "Inter - Domain Routing protocol: EGP, BGP, and IDRP", in Prentice hall,1995.
[5] L.Breslau and D.Estrin, "Design of Inter-Administrative Domain routing protocols", Proceedings of SIGCOMM' 90, pp. 231-241, September 1990.
[6] R.Perlman and G.Varghese, "Pitfalls in the Design of Distributed routing algorithms", Proceediings of SIGCOMM' 88, pp. 43-54, August 1988.
[7] R.Perlman, "A comparison between two routing protocols: OSPF and IS-IS", IEEE Network, 5(5), pp. 18-24, September 1991.
[8] C.Baransel, W.Dobosiewicz, and P.Gburzynski, "Routing in multihop packet switching networks", IEEE Network, 9(3), pp. 38-61, May/June 1995.
[9] Vem Paxson, "End to End Routing behavior in the internet", Lawrence Berkeley National Laboratory, University of California, Berkeley, May 1996.
[10] B.Chinoy, "Dynamics of Internet routing information", Proceedings of SIGCOMM' 93, pp.45-52, September 1993.
[11] S. Deering and D. Cheriton, "Multicast Routing in datagram internetworks and extended LANS", ACM Transactions on Computer Systems, 8 (2), pp. 85 - 110, May 1990.
[12] W.R.Stevens and Comer, TCP/IP Illustrated, volume 1: The Protocols, Addison Wesley, 1994.
[13] William Stallings, High Speed Networks - TCP/IP and ATM design Principles,Prentice hall, 1998.
[14] S.Keshav, An Engineering approach to Computer Networking - ATM networks, the internet, Addison wesley, 1997.
[15] Mark Dickie, Routing in today's Internetworks, VNR Communications laboratory,1993.
[16] John D. Spragins with Joseph L. Hammond and Krzysztof Pawlikowski, Telecommunications Protocols and Design, Addison-Wesley, July 1992.
[17] Christian Huitema, Routing in the Internet, Prentice - Hall, 1995.

PERFORMANCE OF TELECOMMUNICATION SYSTEMS: SELECTED TOPICS

K. Kontovasilis
N.C.S.R. "Demokritos", Inst. Informatics and Telecommunications,
P.O. Box 60228, GR-15310 Aghia Paraskevi Attikis, Greece

S. Wittevrongel, H. Bruneel
Ghent University, Dept. Telecommunications and Information Processing
SMACS Research Group, St. Petersnieuwstraat 41, B-9000 Gent, Belgium

B. Van Houdt, C. Blondia
University of Antwerp, Dept. Mathematics and Computer Science
PATS Research Group, Universiteitsplein 1, B-2610 Antwerpen, Belgium

1. Introduction

The central activity of performance evaluation is building formal descriptions of the system under study, an activity refered to as modelling. These models include workload models (e.g. packet traffic models), system resource models (e.g. switch models, link models) and resource control mechanism models (e.g. MAC protocol models). They are used to gain insight in the performance of the system under certain load conditions. To obtain the performance measures of interest, two technics exist: simulate the system (i.e. built a program that simulates the model behavior) or solve the model mathematically (i.e. compute the performance measures analytically). In this paper, we concentrate on the latter, making distiction between analytical methods that lead to closed formulas (as described in the part on generating functions) and algorithms that allow to compute the measures numerically (as in dealing with matrix analytic methods).

An area where performance modeling is an essential tool for system designers and developers today is the Internet. The Internet is evolving from a best-effort network towards a system that combines Quality of

Service (QoS) support with efficient resource usage. The extremely rapid pace of change that can be observed in the Internet research community (e.g. in the IETF), often does not allow rigorous performance evaluation of the different proposals. Therefore, the performance evaluation community (e.g. IFIP WG 6.3) should make an effort to provide the necessary methods, techniques and tools to allow a better insight into the system behavior.

The first part is devoted to the modeling of telecommunications sytems using generating functions, the second part to modeling using matrix analytical methods and finaly, the third part to the use of asymptotic approximations. The first part considers discrete-time queueing models as representatin of telecommunication systems. These models are particularly applicable when the time can be segmented in intervals of fixed lenght (called slots) and information packets are transmitted at slot boundaries. A typical example is an ATM transmission system. The method to compute the performance measures of interest is based on the use of generating functions. The aim is to obtain a closed form formula for the generating function of the system content (i.e. how many packets are present in the system), from which the most important performance measures can be derived. The main characteristic of this approach is that it is almost entirely analytical.

A second approach to compute performance measures is found in Section 3. Here two recent and promising developments within the framework of matrix analytical methods are discussed. Both models, have important applications in the performance analysis of telecommunication systems. The first model is concerned with a markovian arrival process with marked arrivals, of particular interest in systems where the packets are originating from different possiblly correlated traffic streams. A second model deals with Tree structured Markov chains. Their particular structure can be exploited to obtian efficient computational methods to obtain the measures of interest. Random access algorithms known as stack algorithms, or tree algorithms with free access, are examples of systems that can be modeled by means of tree stuctured Markov chain, leading to expressions for the maximum stable throughput and mean delay in such systems.

A third part is devoted to asymptotic approximations. These methods have become extremely relevant due to the high transmission rates and the stringent quality of service guarantees of modern systems, which make very rate events (e.g. buffer overflow) significant. Hence, performance measures are based on distribution tails, rather than on first moments. This paper studies both asymptotics for multiplexers with small buffers and for multiplexers with large buffers. In both cases, the aim is to link the traffic load and resource capacity to the probability of loss due to buffer overflow. Also the case where the buffer space is neither negligible nor dominant is discussed.

2. Performance Modeling of Communication Systems Using Generating Functions

2.1. Discrete-time queueing models

In various subsystems of telecommunication networks, buffers are used for the temporary storage of digital information units which cannot be transmitted to their destination immediately. The performance of a communication network may be very closely related to the behavior of these buffers. For instance, information units may get lost whenever a buffer is fully occupied at the time of their arrival to this buffer, they may experience undesirable delays or delay variations in buffers, ... Queueing theory thus plays an important role in the performance modeling and evaluation of telecommunication systems and networks. In particular, queueing models in discrete time are very appropriate to describe traffic and congestion phenomena in digital communication systems, since these models reflect in a natural way the synchronous nature of modern transmission systems, whereby time is segmented into intervals ("slots") of fixed length and information packets are transmitted at slot boundaries only, i.e., at a discrete sequence of time points.

In a discrete-time queueing model, the arrival stream of digital information into a buffer (the input or arrival process) is commonly characterized by specifying the numbers of arriving packets during the consecutive slots. In basic models, these numbers of arrivals are assumed to be independent and identically distributed (i.i.d.) discrete random variables, and the corresponding arrival process is referred to as an independent or uncorrelated arrival process. More advanced models allow the numbers of arrivals during consecutive slots to be nonindependent, and are referred to as correlated arrival processes. The storage capacity of a buffer is usually modeled as unlimited. This is an acceptable assumption since in most communication systems the capacity is chosen in such a way that the loss probabilities are very small, and furthermore, this facilitates the use of analytical analysis techniques. The transmission of information units from the buffer (the output process) is characterized by the distribution of the transmission times of the information units, the number of output channels of the buffer, the availability of the output channels, and the order of transmission (the queueing discipline). In basic models, all information units are assumed to be of fixed length, which implies they have constant transmission times, the output channels are permanently available, and the queueing discipline is assumed to be first-come-first-served (FCFS). In some applications, however, it is necessary to consider non-deterministic transmission times, interruptions of the output channels, and non-FCFS queueing disciplines such as e.g. priority queueing.

In the next section, we present an overview of a number of fundamental techniques for the analysis - in the steady state - of a wide range of discrete-time queueing models. The main characteristics of the techniques are that they are almost entirely analytical (except for a few minor numerical calculations) and that an extensive use of probability generating functions is being made. Note that a steady state only exists if the mean number of packet arrivals per slot is strictly less than the mean number of packets that can be transmitted per slot.

2.2. Steady-state queueing analysis using generating functions

The behavior of a queueing system is commonly analyzed in terms of the probability distributions of the buffer contents, i.e., the total number of packets present in the buffer system, and the packet delay, i.e., the amount of slots a packet spends in the system.

Buffer contents. The first step in the analysis of the buffer contents is to establish a so-called "system equation" that describes the evolution in time of the buffer contents. If we define s_k as the buffer contents at the beginning of slot k, it is easily seen that the following basic relationship holds :

$$s_{k+1} = s_k - t_k + e_k , \qquad (1)$$

where e_k represents the total number of packet arrivals during slot k and t_k denotes the number of packets that leave the buffer system at the end of slot k. Here the characteristics of e_k depend on the specific nature of the arrival process. The random variable t_k on the other hand depends on the characteristics of the output process, and cannot be larger than s_k in view of the synchronous transmission mode, which implies that only those packets present in the buffer at the beginning of a slot are eligible for transmission during the slot.

In the simplest models, uncorrelated arrivals from slot to slot, constant transmission times of one slot each, and permanently available output channels are assumed. In this case, the system equation (1) reduces to

$$s_{k+1} = (s_k - c)^+ + e_k . \qquad (2)$$

Here $(...)^+ = max(0,...)$, c denotes the number of output channels, and the random variables s_k and e_k on the right-hand side are statistically independent of each other, which implies that the set $\{s_k\}$ forms a Markov chain. Let $S_k(z) = E[z^{s_k}]$ denote the probability generating function (pgf) of s_k. By means of standard z-transform techniques [9], the system equation (2) can then be translated into the z-domain. This yields the following relationship between the pgf's $S_{k+1}(z)$ and $S_k(z)$:

$$S_{k+1}(z) = E(z)\, z^{-c} \{\sum_{j=0}^{c-1}(z^c - z^j)\operatorname{Prob}[s_k = j] + S_k(z)\}\ , \qquad (3)$$

where $E(z)$ denotes the pgf of the number of packet arrivals in a slot. In the steady state, both $S_{k+1}(z)$ and $S_k(z)$ will converge to a common limiting function $S(z)$, the pgf of the buffer contents s as the beginning of an arbitrary slot in the steady state. Taking limits for $k \to \infty$ and solving the resulting equation for $S(z)$, we then obtain

$$S(z) = \frac{E(z) \sum_{j=0}^{c-1}(z^c - z^j)\operatorname{Prob}[s = j]}{z^c - E(z)}\ . \qquad (4)$$

The c unknown constants $\operatorname{Prob}[s = j]$, $0 \le j \le c-1$, in (4) can be determined by invoking the analyticity of the pgf $S(z)$ inside the unit disk $\{z : |z| \le 1\}$ of the complex z-plane, which implies that any zero of the denominator of (4) in this area must necessarily also be a zero of the numerator, together with the normalization condition $S(1) = 1$ of the buffer-contents distribution. This results in a set of c linear equations in the c unknown probabilities and allows to obtain $S(z)$ explicitly.

During the last few years research has largely focused on the introduction of more complicated characterizations of the arrival process, in order to obtain more realistic, useful and tractable stochastic descriptions of the sometimes bursty and heterogeneous traffic streams occurring in modern integrated communication networks. When the arrival process is correlated, the random variables s_k and e_k on the right-hand side of the system equation (2) are no longer statistically independent, and the above analysis technique needs to be modified. Specifically, since the knowledge of the value of s_k no longer suffices to determine the probability distribution of s_{k+1}, the set $\{s_k\}$ does no longer form a Markov chain, and a more-dimensional state description of the system has to be used, containing extra information about the state of the arrival process.

As an example, let us consider a discrete-time queueing model with one output channel, that is permanently available, and a simple correlated arrival process. Packets are generated by N independent and identical on/off-sources. Each source alternates between on-periods, during which it generates one packet per slot, and off-periods, during which no packets are generated. The successive on-periods and off-periods of a source are assumed to be independent and geometrically distributed with parameters α and β respectively. Clearly, we then have the system equation (2), where $c = 1$, whereas e_k can be derived from e_{k-1} as follows [7]:

$$e_k = \sum_{i=1}^{e_{k-1}} c_i + \sum_{i=1}^{N-e_{k-1}} d_i \ . \tag{5}$$

Here the c_i's and the d_i's are two independent sets of i.i.d. Bernoulli random variables with pgf's

$$c(z) = 1 - \alpha + \alpha z \tag{6}$$

and

$$d(z) = \beta + (1 - \beta) z \ . \tag{7}$$

From (2) and (5)-(7), the pair (e_{k-1}, s_k) is easily seen to constitute a (two-dimensional) Markovian state description of the system at the beginning of slot k. We then define $P_k(x, z)$ as the joint pgf of the state vector (e_{k-1}, s_k), i.e.,

$$P_k(x, z) = E[x^{e_{k-1}} z^{s_k}] \ . \tag{8}$$

The next step is then similar to the uncorrelated-arrivals case, namely to derive a relationship between the pgf's $P_{k+1}(x, z)$ and $P_k(x, z)$ corresponding to consecutive slots, by means of the state equations:

$$\begin{aligned} P_{k+1}(x, z) &= E\left[(xz)^{e_k} z^{(s_k-1)^+}\right] \\ &= [d(xz)]^N E\left[\left(\frac{c(xz)}{d(xz)}\right)^{e_{k-1}} z^{(s_k-1)^+}\right] \\ &= \frac{[d(xz)]^N}{z} \left\{P_k\left(\frac{c(xz)}{d(xz)}, z\right) + (z-1)\operatorname{Prob}[s_k = 0]\right\} \ . \end{aligned} \tag{9}$$

Again taking limits for $k \to \infty$, we now obtain a "functional equation" for the limiting function $P(x, z)$, which typically contains the P-function on both sides, but with different arguments. Although the function $P(x, z)$ cannot be derived explicitly from the functional equation, several performance measures related to the buffer contents can be derived from it, as will be explained later.

A similar analysis is possible for a variety of correlated arrival processes, such as train arrivals [67], [65], Markov modulated arrivals [68],

[2], general on/off sources [64], correlated train arrivals [47], and so on. For some arrival processes, the resulting functional equation may contain a number of unknown boundary probabilities, which in general are difficult to obtain exactly. An approximation technique can then used, which is based on the observation that a buffer contents equal to n at the beginning of a slot implies that no more than n packets have entered the buffer during the previous slot (see e.g. [2], [64]).

Also in case more complicated models for the output process are used, similar problems occur and a more-dimensional state description needs to be used. For instance, when general transmission times are considered, additional information is needed in the state description about the amount of service already received by the packet(s) in transmission, if any [8]. In case interruptions of the output channels may occur, we need to keep track of the state of each of the output channels (available or blocked) and the remaining sojourn time in this state [15].

Packet delay. The delay of a packet is defined as the number of slots between the end of the slot of arrival of the packet, and the end of the slot when this packet leaves the buffer. In case of a FCFS queueing discipline, the analysis of the packet delay typically involves the derivation of a relationship between the delay of a tagged packet and the total number of packets present in the buffer just after the arrival slot of the tagged packet and to be transmitted before the tagged packet. However, for discrete-time queueing systems with one permanently available output channel, constant transmission times of one slot, a FCFS queueing discipline and an arbitrary (possibly correlated) arrival process, the following relationship exists between the pgf $S(z)$ of the buffer contents and the pgf $D(z)$ of the packet delay [59]:

$$D(z) = \frac{S(z) - S(0)}{1 - S(0)} . \qquad (10)$$

The above relationship makes a full delay analysis superfluous, once the buffer contents has been analyzed. Similar relationships also exist in case of multiple servers [61] and non-deterministic service times [60].

2.3. Performance measures

The results of the analysis can be used to derive simple and accurate (exact or approximate) formulas for a wide variety of performance measures of practical importance, such as mean and variance of buffer occupancies and delays, packet loss probabilities, ... The mean system contents and the mean packet delay in the steady state can be found by evaluating the first derivative of $S(z)$ and $D(z)$ at $z = 1$. Higher-order moments of the system contents and the packet delay can be derived analogously, by calculating higher-order derivatives of $S(z)$ and $D(z)$ at $z = 1$. The tail distribution of the buffer contents is, for reasons of computational complexity, often approximated by a geometric form based on the dominant pole z_0 of the pgf of the buffer contents. That is, for large values of n, the tail distribution of the buffer contents is approximated by [26]

$$\text{Prob}[s = n] \approx -\frac{\theta}{z_0} z_0^{-n}, \qquad (11)$$

where θ is the residue of $S(z)$ for $z = z_0$. A quantity of considerable practical interest is the probability that the buffer contents (in the infinite buffer) exceeds a given threshold S. This probability can be used to derive an approximation for the packet loss ratio (i.e., the fraction of packets that arrive at the buffer but cannot be accepted) of a buffer with finite waiting space S and the same arrival statistics [50].

As mentioned before, it is not always possible to calculate the pgf $S(z)$ of the buffer contents explicitly. Nevertheless, a technique has been developed to derive results concerning the moments and the tail distribution of the buffer contents from the associated functional equation. The technique involves considering those values for which the first argument(s) of the $P(.,z)$ functions in both sides of the functional equation become equal (see e.g. [7], [68], [64], [66]).

2.4. Numerical example

As an illustration, we consider a statistical multiplexer to which messages consisting of a variable number of fixed-length packets arrive at the rate of one packet per slot ("train arrivals"), which results in a *primary* correlation in the packet arrival process. The arrival process contains an additional *secondary* correlation, which results from the fact that the distribution of the number of leading packet arrivals (of new messages) in a slot depends on some environment variable. This environment has two possible states 'A' and 'B', each with geometrically distributed sojourn times [47]. We compare the results obtained for this correlated train arrivals model with the results that would be found if a model without secondary correlation or an uncorrelated model for the packet arrival process were used. In Figure 1, the mean buffer contents for the three considered arrival models, i.e., $E[s]$ (correlated train arrivals), $E[s_{\text{prim}}]$ (uncorrelated train arrivals) and $E[s_{\text{un}}]$ (uncorrelated packet arrivals) are plotted versus the total load ρ, for different values of the environment correlation factor K, which can be seen as a measure for the absolute lengths of the sojourn times, when their relative lengths are

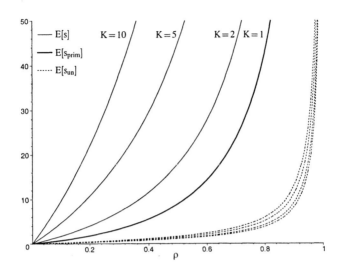

Figure 1. Mean buffer contents versus the total load ρ for various values of K.

given [47]. The message-length distribution is a mixture of two geometrics according to the pgf $L(z) = \frac{0.5(1-\lambda)z}{1-\lambda z} + \frac{0.5(1-\tau)z}{1-\tau z}$ with mean 5 and a variance of 50. In an 'A'-slot, the number of new messages has a geometric distribution with mean 2, while no new messages are generated during 'B'-slots. The figure clearly shows the severe underestimation of the buffer contents when the different levels of correlation in the arrival process are neglected. Note that all the curves for $E[s_{\text{prim}}]$ coincide with the one representing $E[s]$ for $K = 1$ (uncorrelated environment). In the case of uncorrelated packet arrivals, $E[s_{\text{un}}]$ slightly increases with higher values of K, although not in the same drastic way as $E[s]$ in case of correlated train arrivals.

3. Performance Modeling using Matrix Analytic Methods

Two recent, and promising, developments within the area of matrix analytic methods are discussed in this section. It concerns the Markovian arrival process with marked arrivals, i.e., the MMAP[K] arrival process, and tree structured Markov chains of the $M/G/1$, $GI/M/1$ and Quasi-Birth-Death (QBD) type. While presenting these new developments, we mainly focus on their applicability towards telecommunication systems.

Matrix analytic methods, for queueing theory, found their origin in the 1960s in the work of Cinlar and Neuts [17]. During the 1970s, Neuts made a number of crucial contributions to the $M/G/1$ and $GI/M/1$ structures and wrote a book, the use of which is still widespread nowadays, on this subject [45]. During the 1980s, Neuts pursued his work at the University of Delaware together with his associates and students Chakravarthy, Kumar, Latouche, Lucantoni and Ramaswami. In 1989, a second, perhaps somewhat less accessible to computer scientists, book [46] appeared on the $M/G/1$ structure that summarizes their achievements, it reflects the fact that the area of matrix analytic methods was growing vigorously. The theory and its applications have grown unabated ever since. This was clearly demonstrated in 1996, when the first conference on matrix analytic methods, and its applications, was organized. At the time of this writing a fourth conference will be held in July 2002 in Adelaide, Australia.

3.1. Markovian Arrival Process with Marked Arrivals

The usefulness of queueing theory as a means of analyzing the performance of telecommunication systems has been demonstrated extensively. However, until recently, most of the work done in this area applied to queueing systems that do not distinguish between customers, that is, all customers are of the same type and require the same type of service. There are plenty of applications were it would be suitable to distinguish between multiple customer types. For example, suppose that packets originating from K different, possibly correlated, traffic streams form the input to a buffer, then it is often useful if we could obtain statistics, e.g., the delay distribution, for each individual source. The Markovian arrival process with marked arrivals, i.e., the MMAP[K] process, is an important building block that allows us to obtain such information.

Both a continues and a discrete time version of the MMAP[K] arrival process has been introduced [31, 27], but we restrict ourselves to the discrete time variant. We shall distinguish between two types of MMAP[K] processes: those that allow for batch arrivals to occur, and those that do not.

MMAP[K] Process without Batch Arrivals. A discrete time MMAP[K] arrival process that does not allow for batches to occur is a natural extension of the D-MAP arrival process [5]. Customers are distinguished into K different types. The MMAP[K] is characterized by a set of $m \times m$ matrices $\{D_k \mid 0 \leq k \leq K\}$, with m a positive integer. The $(j_1, j_2)^{th}$ entry of the matrix D_k, for $k > 0$, represents the probability that a type k customer arrives and the underlying Markov chain (MC) makes a transition from state j_1 to state j_2. The matrix D_0

covers the case when there are no arrivals. The matrix D, defined as

$$D = \sum_{k=0}^{K} D_k,$$

represents the stochastic $m \times m$ transition matrix of the underlying MC of the arrival process. Let θ be the stationary probability vector of D, that is, $\theta D = \theta$ and $\theta e = 1$, where e is a column vector with all entries equal to one. The stationary arrival rate of type k customers is given by $\lambda_k = \theta D_k e$. Queues with MMAP[K] arrival processes are discussed in Section 3.1

Example 3.1. Consider a D-MAP arrival process characterized by the $m \times m$ matrices $\tilde{C}$ and $\tilde{D}$. Suppose that we wish to mark the arrivals by the state of the underlying MC at its generation time. This results in a MMAP[K] arrival process with $D_0 = \tilde{C}$ and with the matrices D_k, for $1 \le k \le K = m$, equal to zero, except for their k-th row, which is identical to the k-th row of $\tilde{D}$. Notice, the number of customer types might be smaller than m, because some rows of $\tilde{D}$ might be equal to zero.

MMAP[K] Process with Batch Arrivals. A discrete time MMAP[K] arrival process that allows for batches to occur —a natural extension of the D-BMAP arrival process [5]—is characterized by a set of $m \times m$ matrices D_C where C is a string of integers between 1 and K, that is, $C = c_1 \ldots c_b$ with $1 \le c_l \le K$ and $1 \le l \le b$. Let b_{max} be the maximum batch size of the MMAP[K] arrival process. Let $\emptyset$ denote the empty string and $|C|$ the length of the string C. The $(j_1, j_2)^{th}$ entry of the matrix D_C, with $C \ne \emptyset$, represents the probability that a batch of $|C|$ arrivals occurs, while the underlying MC makes a transition from state j_1 to state j_2. The type of the l-th customer of the batch is c_l, for $1 \le l \le |C|$, if $C = c_1 \ldots c_{|C|}$. As before, $D = \sum_C D_C$ represent the transition matrix of the underlying MC and θ its stationary probability vector. The stationary arrival rate of type k customers is given by $\lambda_k = \theta \sum_C N(C, k) D_C e$, where $N(C, k)$ counts the number of occurrences of the integer k in the string C. Queues with MMAP[K] arrival processes are discussed in Section 3.1

Example 3.2. It is well known that a superposition of two, or more, D-BMAPs is again a D-BMAP. However, when superposing D-BMAPs customers generally loose their identity, meaning that we no longer know whether the arrival came from the first or the second D-BMAP. A MMAP[K] arrival process that eliminates this drawback can be constructed in the following way. Suppose that the first, resp. second, D-BMAP is characterized by the $m_1 \times m_1$ matrices $\tilde{D}_n^1$, resp. $m_2 \times m_2$ matrices $\tilde{D}_n^2$, for $n \geq 0$. Let D_C, with C a string of $b_1 \geq 0$ ones followed by $b_2 \geq 0$ twos[1], be $m_1 m_2 \times m_1 m_2$ matrices. Instead of labeling the $m_1 m_2$ states j of the underlying MC as 1 to $m_1 m_1$, we denote them as (j, j'), with $1 \leq j \leq m_1$ and $1 \leq j' \leq m_2$. The $(\mathbf{j_1}, \mathbf{j_2})^{th}$ entry, with $\mathbf{j_1} = (j_1, j_1')$ and $\mathbf{j_2} = (j_2, j_2')$, of the matrix D_C, with C a string of b_1 ones followed by b_2 twos, equals $(\tilde{D}_{b_1}^1)_{j_1, j_2} (\tilde{D}_{b_2}^2)_{j_1', j_2'}$. A variety of examples is presented in [31, 29].

The MMAP[K]/PH[K]/1 Queue. In this section we discuss the MMAP[K]/PH[K]/1 queue with a first-come-first-serve (FCFS) and a last-come-first-serve (LCFS) service discipline. The service times of type k customers, in a MMAP[K]/PH[K]/1 queue, have a common phase-type distribution function with a matrix representation (m_k, α_k, T_k), where m_k is a positive integer, α_k is an $1 \times m_k$ nonnegative stochastic vector and T_k is an $m_k \times m_k$ substochastic matrix. Let $T_k^0 = e - T_k e$, then the mean service time of a type k customer equals $1/\mu_k = \alpha_k (I - T_k)^{-1} e$. The i-th entry of α_k represents the probability that a type k customer starts its service in phase i. The i-th entry of T_k^0, on the other hand, represents the probability that a type k customer completes its service provided that the service process is in phase i, while the (i, j)-th entry of T_k equals the probability that it does not complete its service and the phase at the next time instance is j.

The positive recurrence, i.e., stability, of these queues was studied by He in [28]. Explicit formulas for the Laplace-Stieltjes transforms of the waiting times of a type k customer have been obtained for a server with a FCFS service discipline [29]. An algorithm to obtain the steady state probabilities of a MMAP[K]/PH[K]/1 queue, where the MMAP[K] arrival process does not allow for batches to occur and the server follows a LCFS service discipline, is found in [30]. Finally, a simple algorithm, based on the $GI/M/1$ structure, has been developed to calculate the delay distribution of a type k customer in a FCFS MMAP[K]/PH[K]/1 queue [54]. This algorithm is highly efficient if the MMAP[K] arrival process does not allow for large batch arrivals to occur.

Example 3.3. Let us continue with the MMAP[2] arrival process introduced in Example 3.2. Now, assume that each of the two D-BMAPs model a traffic source and that the traffic generated by both sources

---

[1]For simplicity, we assume that the arrivals of the first D-BMAP occur before those of the second, there is however no need to do so.

share a buffer. Moreover, assume that the packets generated by source k, for $k = 1,2$, have a fixed length of L_k bytes. Then, this buffer can be modeled by a discrete time MMAP[2]/PH[2]/1 queue, because fixed length service times have a phase type distribution. As a result, we could calculate the delay distribution of a source k arrival using [54].

Example 3.4. Many random access algorithms (RAAs) that use grouped access as their channel access protocol (CAP) can be modeled in a natural way by means of a MMAP[K]/PH[K]/1 queue ([52, 55, 56]). When modeling such a RAA, a type k customer corresponds to a group of k contenders and its service time distribution is the time necessary for each of the k contenders to successfully transmit their packet, starting from the completion time of the previous group.

3.2. Tree Structured Markov Chains

Another promising development in the theory of matrix analytic methods are tree structured Markov chains (MCs). Consider a discrete time bivariate MC $\{(X_t, N_t), t \geq 0\}$ in which the values of X_t are represented by nodes of a d-ary tree, and where N_t takes integer values between 1 and m. X_t is referred to as the node and N_t as the auxiliary variable of the MC at time t. A d-ary tree is a tree for which each node has d children. The root node is denoted as $\emptyset$. The remaining nodes are denoted as strings of integers, with each integer between 1 and d. For instance, the k-th child of the root node is represented by k, the l-th child of the node k is represented by kl, and so on. Throughout this paper we use lower case letters to represent integers and upper case letters to represent strings of integers when referring to nodes of the tree. We use '+' to denote concatenation on the right, e.g., if $J = j_1\ j_2\ j_3, k = j$ then $J + k = j_1\ j_2\ j_3\ j$. If J can be written as $K_1 + K_2$ for some strings K_1 and K_2, K_1 is called *an ancestor* of J.

Algorithms that allow for the calculation of the steady state probabilities, have been identified for three subsets of the tree structured MCs, each subset allows for a certain type of transitions to occur:

- The *skip-free to the left*, i.e., M/G/1 Type, MCs: It is impossible to move from node J to $\emptyset$, without visiting *all ancestors* of J [51].

- The *skip-free to the right*, i.e., GI/M/1 Type, MCs: Transitions from a node J are allowed to the root node $\emptyset$, the *children* of J and the *children of all ancestors* of J [70].

- The Quasi-Birth-Death (QBD) MCs: The chain can only make transitions to its parent, children of its parent, or to its children [69].

So far, the last subset has proven to be the most fruitful. Therefore, they are discussed in more detail in this section. If a tree structured QBD MC is in state $(J+k,i)$ at time t then the state at time $t+1$ is determined as follows:

1. (J,j) with probability $d_k^{i,j}, k = 1, \ldots, d,$

2. $(J+s,j)$ with probability $a_{k,s}^{i,j}, k, s = 1, \ldots, d,$

3. $(J+ks,j)$ with probability $u_s^{i,j}, s = 1, \ldots, d.$

Define $m \times m$ matrices $D_k, A_{k,s}$ and U_s with respective $(i,j)^{th}$ elements given by $d_k^{i,j}, a_{k,s}^{i,j}$ and $u_s^{i,j}$. Notice that transitions from state $(J+k,i)$ do not dependent upon J, moreover, transitions to state $(J+ks,j)$ are also independent of k. When the Markov chain is in the root state $(J = \emptyset)$ at time t then the state at time $t+1$ is determined as follows:

1. $(\emptyset, j)$ with probability $f^{i,j}$,

2. (k,j) with probability $u_k^{i,j}, k = 1, \ldots, d.$

Define the $m \times m$ matrix F with corresponding $(i,j)^{th}$ element given by $f^{i,j}$. Algorithms that calculate the steady state probabilities using the matrices $D_k, A_{k,s}, U_s$ and F as input parameters are available in [69, 4].

Example 3.5. MMAP[K]/PH[K]/1 queue, where the MMAP[K] arrival process does not allow for batches to occur, with a last-come-first-serve (LCFS) service discipline can be modeled using a tree stuctured QBD MC [30]. Indeed, the line of customers waiting in a MMAP[K]/PH[K]/1 queue can be represented by a string of integers between 1 and K, thus as nodes of a K-ary tree. The auxiliary variable is used to represent the phase of the server, the type of customer in the server and the state of the MMAP[K] arrival process. The root node $\emptyset$ corresponds to a queue with a busy server and an empty waiting room. Therefore, one needs a generalized boundary condition to represent the situation where the waiting room is empty and the server is not busy. Information on generalized boundary conditions and other extension, i.e., MCs with a *forrest* structure, can be found in [70].

Example 3.6. Random access algorithms (RAAs) known as stack algorithms, or tree algorithms with free access, can be modeled using a tree structured QBD MC [53, 57]. As a result, it is possible to study the maximum stable throughput, as well as the mean delay, for various D-BMAP (and BMAP) arrival processes.

4. Asymptotic approximations for the performance evaluation of large broadband networks

4.1. The need for asymptotic methods

After a period of intensive development, multiservice broadband networks are now a reality. Current implementations already serve as high-speed backbone infrastructures and more extensive usage, accompanied by a further exploitation of these networks' advanced capabilities, is expected when the need for providing complex information services with strict quality guarantees will grow.

There are two primary performance-related characteristics that distinguish multiservice broadband networks from their "conventional" counterparts. The first is that, due to both the high transmission speed and the need for providing individualized—and stringent—quality of service (QoS) guarantees, very rare events (e.g., buffer overflows occurring with probability as low as 10^{-6}, or smaller) become significant. Consequently, most relevant performance metrics must be based on distribution tails rather than mean values. The second characteristic is that most bandwidth-demanding traffic types appearing on broadband networks are bursty, i.e., they feature significant rate excitations and correlated packet interarrival times. These are properties that leave a mark on the queueing phenomena governing the network's performance.

The typical queueing effects of burstiness are demonstrated by the main graph of Fig. 2, depicting the buffer overflow probability (a standard performance metric) at a network multiplexer or switch loaded by a superposition of bursty traffic streams, as a function of the buffer size. Two distinct regions are clearly identified: In the first region (small buffer sizes) the rate correlations do not become apparent, the traffic is primarily characterized (at the, so called, 'packet level') by properties of the individual interarrival times between successive packets, and the overflow probability decays rapidly with increasing buffer size (at an exponentially fast rate, since the graph uses a log-linear scale). In the second region (larger buffer sizes) the rate correlation details (usually collectively called 'burst level traffic properties') become noticeable, resulting in a quite smaller rate of decay for the overflow probability.

Clearly, accurate prediction of tail probabilities—like those in the example—requires the usage of sophisticated traffic models, able of providing a sufficiently precise characterization of traffic at both the packet and burst levels. Such detailed models, and associated analysis methods, do exist and are invaluable whenever thorough queueing analysis is called for. In due account, the paper reviews two important classes

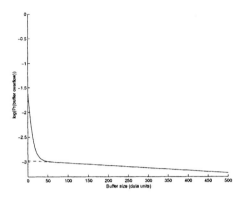

Figure 2. The effect of burstiness: overflow probability vs buffer size (at a log-linear scale).

of models/solution methods (see the sections on matrix analytic techniques, and on generating functions based techniques). Unfortunately, detailed descriptions suffer from the 'state space explosion' problem. Indeed, the state spaces of models for all but the simplest traffic patterns have to be rather large, if both the packet- and burst-level behavior is to be captured. The situation becomes worse when it is realized that, in virtually all congestion phenomena of interest, the aggregate traffic load consists of a (frequently heterogeneous) superposition of a large number of individual streams and that the state space of the model for the aggregate traffic depends factorially on the—already large—spaces of the constituents.

In an attempt to partially alleviate this difficulty, 'fluid-flow' models of traffic have been proposed. These models disregard the discrete nature of the packet level details, representing traffic as the flow of a continuous fluid (hence their name). The instantaneous rate of this flow is taken equal to the average rate of the real traffic over an appropriate time window, large enough to "hide" the packet details, but also small enough to preserve the burst-level rate fluctuations. This approach has been quite successfully employed towards the accurate representation of burst-level traffic dynamics with a reduced set of model parameters. An example is provided by the dashed graph in Fig. 2, which represents the overflow probability curve corresponding to the fluid-flow counterpart of the original traffic and which matches quite satisfactorily with the exact result over the burst level region. For further information on (primarily Markovian) fluid-flow models see: [1, 40, 43, 49] for the basic theoretical foundation and analysis techniques, [3, 37, 42] for embelishments of the theory and efficient computational algorithms, and [38, 44] for multiple-scale phenomena occurring when the traffic possesses burst-level dynamics with a finer structure.

However, although the fluid-flow concept works for reducing the complexity of models for *individual* traffic streams, it cannot alleviate the state space explosion due to superposition. For this reason, many important performance-related network mechanisms, particularly those that must operate within a short time-frame (such as on-line traffic control) or over a combinatorially large domain (e.g., network-wide resource (re)allocation), cannot rely on "classical" queueing techniques, even the fluid-flow ones.

Fortunately, there's still a viable way of addressing the problem, grounded on the fact that modern broadband networks are, in some respects, "large" systems, featuring high link capacities and large switches, and requiring that probabilities of hazardous events (like overflows leading to data losses) be very small (so as to provide reliable QoS guarantees). This setting suits well to the 'Theory of Large Deviations' (TLD), a body of theoretical results and techniques that address systems "scaled up" by a large parameter and examine the circumstances under which associated (scaled) random variables may attain values in a designated set with an exponentially small probability, asymptotically as the scaling parameter approaches infinity. TLD may be used to compute the rate of exponential decay in the probabilities of interest and, moreover, determine the way in which these 'rare events' occur. A comprehensive general treatment of TLD can be found in, e.g., [16], while [10] provides a less formal exposition, explicitly geared towards applications. Reference [63] may be consulted for a brief overview of topics and further references.

Building on the TLD foundations, the very same characteristics that lead to state-space explosion in "conventional" models have been exploited towards the development of asymptotic theories that quantify congestion in broadband networks under bursty load. The purpose of this section is to give an outline of the relevant results. Before embarking on the review, however, it is important to note that, besides analytical tractability, a prime advantage of the asymptotic methods is their potential for conceptual clarity, something crucial for highlighting the effect of fundamental phenomena in explicit terms.

Generically, two such congestion-related phenomena may be identified: The first, frequently called 'multiplexing gain', relates to the fact that (as a consequence of the law of large numbers) aggregation of many independent traffic streams results in smoother compound traffic, reducing the probability with which the aggregate data rate raises above its mean value. As more streams are multiplexed, the amount of bandwidth per stream required to compensate for the rate excitations is reduced (for a given QoS requirement), justifying the name of the phenomenon. In the absence of significant buffering, multiplexing gain is the only mechanism through which QoS may be attained while using less bandwidth than peak-rate. In Fig. 2 this is reflected at the non-negligible probability of overflow even with a zero buffer size. The relevant asymptotic theory is reviewed in Subsection 4.2.

The second fundamental phenomenon relates to another mechanism of controlling rate excitations so as to avoid data losses, that of temporarily storing excessive data into a buffer. The larger the buffering resource, the smaller the capacity requirement for the output port becomes, for a given loss probability. In analogy with multiplexing gain, this bandwidth-savings effect will be called 'buffering gain'. In Fig. 2 it is reflected at the decay of the overflow probability with increasing buffer size, even at the "slow" burst-level region. The asymptotic theory relevant to buffering gain is reviewed in Subsection 4.3.

The two regimes just outlined relate to either no buffer, or a large buffer, so that either the multiplexing gain, or the buffering gain dominate, respectively. In many cases the available buffer is neither negligible nor dominant and both phenomena are noticeable. For this more general setting there is an improved asymptotic theory that can quantify the combined effect of both gain factors, by considering systems where the load and resources (buffer and bandwidth) are proportionally scaled by a large parameter. Elements of this theory are provided in Subsection 4.4.

4.2. Asymptotics for multiplexers with small buffers

Consider a multiplexer (or an output port unit of a switch) featuring a negligibly small buffer and serving traffic through an output link of capacity equal to C. The aggregate traffic loading this system can be described as a stochastic instantaneous-rate[2] process $\{r(t),\ t \in \mathbb{R}\}$, which it is assumed throughout stationary. Tracking just instantaneous rates is adequate, as there is no buffer to "record the past history" of the traffic. In the following, the properties of the instantaneous rate will be described through the respective log-moment generator (also called the 'cumulant generator') $\phi(s) \cong \log \mathbb{E}\, e^{sr(t)}$. As an implication of stationarity, $\phi(s)$ is independent of time.

---

[2] Here we adopt a fluid approach and represent the flow of data as a continuum. However, all results in this section bear obvious analogies with a discrete-time setting, in which $r(t)$ stands for the amount of data contributed during the time-slot indexed by (the now integer) t.

Performance of Telecommunication Systems

At this point it is reminded that the log-moment generator of a random variable (r.v.) is a convex function (actually strictly convex, unless the r.v. is a.s. constant). The set $\{\,s \in \mathbb{R} \mid \phi(s) < \infty\,\}$ is called the generator's 'effective domain'. If $s = 0$ is in the interior of this domain (a mild condition, assumed throughout and satisfied in all cases of practical interest, in particular when the r.v. is bounded—translated to the existence of a finite peak rate in our case), then the generator is an analytic function on the whole interior of its effective domain. By convexity, the derivative $\phi'(s)$ is increasing (strictly increasing if the r.v. is not a.s. constant) and the same may be shown for $\phi(s)/s$. Furthermore, the limits of these functions are related to the extremal values[3] of the corresponding r.v. X as follows:

$$\operatorname{ess\,inf} X = \lim_{s \to -\infty} \phi'(s) = \lim_{s \to -\infty} \frac{\phi(s)}{s} < \lim_{s \to 0} \frac{\phi(s)}{s} = \mathbb{E}\,X \quad (12)$$
$$= \lim_{s \to 0} \phi'(s) < \lim_{s \to +\infty} \frac{\phi(s)}{s} = \lim_{s \to +\infty} \phi'(s) = \operatorname{ess\,sup} X.$$

Since there is no buffer, overflows (and data losses) occur whenever the instantaneous data rate exceeds the system's capacity. We now derive an upper bound to the probability of overflow. Indeed, by a Chebycheff-type argument, for any $s \geq 0$,

$$\Pr\{\,r(t) > C\,\} = \int_{x=C^+}^{\infty} dF_r(x) \leq \int_{x=C^+}^{\infty} e^{s(x-C)}\, dF_r(x)$$
$$\leq \int_{x=0}^{\infty} e^{s(x-C)}\, dF_r(x) = \exp\{\phi(s) - sC\}.$$

By taking logarithms and optimizing over the permissible range of parameters, we obtain

$$\log \Pr\{\,r(t) > C\,\} \leq -\sup_{s \geq 0}\{Cs - \phi(s)\}. \quad (13)$$

This bound is known in the literature as 'Chernoff's bound'. Assuming a stable system (i.e., $C > \mathbb{E}\,r(t)$), the maximum over nonnegative reals coincides with the maximum over the entire real line, i.e.,

$$\forall C > \mathbb{E}\,r(t) \triangleq \bar{r}, \quad \sup_{s \geq 0}\{Cs - \phi(s)\} = \sup_{s \in \mathbb{R}}\{Cs - \phi(s)\} \triangleq I(C), \quad (14)$$

---

[3] The upper extremal value of a r.v. X, called 'essential supremum' and denoted by $\operatorname{ess\,sup} X$, is the largest value that X is not improbable to exceed, namely $\operatorname{ess\,sup} X = \sup\{\,x \in \mathbb{R} \mid \Pr\{X > x\} > 0\,\}$. The lower extremum, called 'essential infimum' and denoted by $\operatorname{ess\,inf} X$, is defined analogously.

the value of the Fenchel-Legendre transform of $\phi(\cdot)$ at C. Furthermore, it may be shown that, for $C > \bar{r}$, the Fenchel-Legendre transform $I(\cdot)$ is an increasing function (actually strictly increasing, unless $r(t)$ is a.s. constant), expressing the intuitively appealing fact that the overflow probability becomes smaller as the system's capacity increases.

Assume now that the aggregate traffic consists of a large number n of independent and identically distributed streams, while the system's capacity is proportionally scaled, maintaining a fixed amount of bandwidth per source, i.e., $C = nc$. Since log-moment generators are additive for independent r.vs, the aggregate generator is $\phi_n(s) = n\phi(s)$ (where now $\phi(\cdot)$ signifies the generator of a single stream) and from equations (13) and (14) it follows that the overflow probability is bounded by $e^{-nI(c)}$, decaying exponentially with large n at a rate equal to $I(c)$. This reflects the fact that, as more sources are multiplexed and the bandwidth per source c remains fixed, overflows become less probable, because the compound traffic "smoothens". In other words, and due to the monotonicity of $I(\cdot)$, a smaller value of c is required as n increases, for a given target overflow probability. This is exactly the multiplexing gain phenomenon, discussed in the previous subsection.

The Chernoff bound of eq. (13) is conservative, allowing for safe performance-related decisions. Not only that, but the bound is asymptotically tight, as the number of sources $n \to \infty$. Specifically, by Cramér's Theorem (see, e.g., [16, Theorem 2.2.3]), it holds

$$\lim_{n \to \infty} \frac{1}{n} \log \Pr\{r_n(t) > nc\} = -I(c), \qquad (15)$$

where, as with the generator, $r_n(t)$ denotes the aggregate rate. This result suggests that, when n is large enough, the probability of overflow is $e^{-\epsilon}$, where $\epsilon = nI(c) + o(n)$. (The quantity ϵ expresses the achievable QoS at a logarithmic scale and will be called the 'quality level' in the sequel.) There is also a more detailed result, called the 'Bahadur-Rao' correction, that strengthens the asymptotic equivalence of (15) to linear, rather than logarithmic order. (In this result, $I(c)$ still remains the dominant factor determining the probability of overflow.) For details, see, e.g., [16, Theorem 3.7.4].

When the traffic is a heterogeneous mix of independent traffic streams, the previous theory still applies. Indeed, consider k traffic classes, each containing n_i, $i = 1, \ldots, k$ independent and identical streams. Then the total number of sources is $n = \sum_{i=1}^{k} n_i$ and the aggregate generator is constructed by the individual counterparts through $\phi_n(s) = \sum_{i=1}^{k} n_i \phi_i(s)$. In this setting (15) still holds, i.e., for large n the probability of overflow is approximately $e^{-\epsilon}$, with quality level $\epsilon = nI(c) = \sup_s \{Cs - \sum_{i=1}^{k} n_i \phi_i(s)\}$.

We now discuss the computation of the decay rate in the asymptotic (15). Due to the convexity of log-moment generators, the function to be maximized in (14) is concave and attains a unique maximum. Moreover, by differentiability (again borrowed from the generator) the derivative of the function in (14) is zero at the maximizing argument. From these observations and from eq. (12) it follows that when the capacity C is between the aggregate mean and peak rates, the quality level is computed as

$$\epsilon = nI(C/n) = \sup_{s \geq 0}\{Cs - \sum_{i=1}^{k} n_i \phi_i(s)\} = Cs^* - \sum_{i=1}^{k} n_i \phi_i(s^*), \quad (16)$$

where s^* is the unique[4] argument satisfying

$$\sum_{i=1}^{k} n_i \phi'_i(s^*) = C, \quad (17)$$

and where the equations have been expressed in a form suitable for a general heterogeneous traffic mix.

Usually, (16) and (17) must be solved numerically. However, the canonical example of a homogeneous on/off traffic mix admits a closed form solution. Indeed, for any on and off sojourn distributions (just assuming finite means, respectively $E T_{\text{on}}$ and $E T_{\text{off}}$) each constituent rate process is stationary and ergodic. By letting $p = E T_{\text{on}}/(E T_{\text{on}} + E T_{\text{off}})$ stand for the probability of visiting the on-state, the instantaneous rate of a single stream is Bernoulli distributed, with generator $\phi(s) = \log[pe^{sr} + (1-p)]$, where r is the stream's peak rate. Then, application of (17) and (16), yields

$$\epsilon = n\left(\beta \log \frac{\beta}{p} + (1-\beta)\log \frac{1-\beta}{1-p}\right), \quad \text{where} \quad p < \beta \triangleq \frac{C}{nr} < 1.$$

Up to this point, the focus of the discussion was on estimating the system's performance under given resources and traffic load. However, network traffic engineering usually deals with problems of an "inverse" nature. One particularly important one is the so called, traffic admission control (also named connection admission control—CAC), where the network resources (in our case the multiplexer's capacity C) and the

---

[4]Except in the trivial case where all traffic rates are a.s. constant. This case is excluded here.

desired quality level ϵ are given and the task consists of deciding whether a candidate traffic mix may be admitted by the network while still satisfying the QoS requirement. Formally, assume that the traffic load at a multiplexer may consist of a superposition of streams from k different traffic classes, each with known characteristics (quantified through the respective generators $\phi_i(\cdot)$, $i = 1, \ldots, k$) and let a potential traffic mix be represented by the vector $\boldsymbol{n} = (n_1, \ldots, n_k)$, with elements the numbers of streams from each class participating in the mix. In this notation, a traffic mix may be admitted without violating the QoS, iff it belongs to the so called admission domain $\{\boldsymbol{n} \mid f(\boldsymbol{n}) \geq \epsilon\}$, where $f(\boldsymbol{n})$ stands for the right hand side of (16).

Given this framework, traffic admission control could in principle proceed by computing $f(\boldsymbol{n})$ through (17) and (16) and comparing the result to the target quality level ϵ. However, the relevant computations involve *all* traffic classes in the mix, making it difficult to take incremental decisions (useful in the common case when a single new connection asks to join an already accepted—potentially large—mix). For this reason, alternative algorithms are required, which usually rely on determining the boundary of the admission domain (i.e., mixes satisfying $f(\boldsymbol{n}) = \epsilon$). If that boundary was linear, then a particularly simple algorithm would be possible, because there would be constants a_i, $i = 1, \ldots, k$ and b (possibly dependent on C and ϵ but not on the traffic mix), such that the admission domain would contain exactly those $\boldsymbol{n}$ satisfying

$$\sum_{i=1}^{k} a_i n_i \leq b. \tag{18}$$

Thus, for the purposes of admission control, each traffic stream would be completely characterized by the quantity a_i corresponding to its class and incremental admission control would proceed by merely adding this quantity to a register (maintaining the sum for the already present traffic) and comparing to b.

Unfortunately, the boundary of the admission domain, as defined by (16) and (17) is not linear[5]; the typical form of its shape is displayed on Fig. 3 for two traffic classes (ignore for the moment the linear segment). Despite this difficulty, it is still possible to obtain a locally optimal linearization, by observing that, due to (16), $f(\boldsymbol{n})$ is convex and the same holds for the complement of the admission domain. Thus, it is assured that the tangent hyperplane at a point $\boldsymbol{n}^*$ on the boundary will

---

[5] Nonlinearity is unavoidable if the nature of the multiplexing gain phenomenon is to be preserved. This point will be discussed to a greater extent in Subsection 4.3.

Performance of Telecommunication Systems

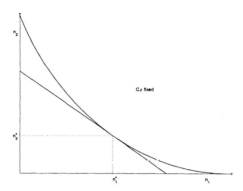

Figure 3. Admission domain for two traffic classes and linear approximation of the boundary around n^*

rest inside the admission domain (see Fig. 3), while also coinciding with the true boundary at n^*. By observing that $f(n^*) = \epsilon$ and by using (16) and (17), it follows that $\partial f/\partial n_i|_{n=n^*} = -\phi_i(s^*(n^*))$ and, further, that the subset of the domain bounded by the tangent hyperplane contains those traffic mixes n satisfying

$$\sum_{i=1}^{k} n_i \frac{\phi_i(s^*(n^*))}{s(n^*)} \leq C - \frac{\epsilon}{s^*(n^*)}. \tag{19}$$

In order to use (19), one must determine a traffic mix n^* at the boundary of the true admission domain and then compute the corresponding value of the maximizing s-parameter, namely $s^*(n^*)$. Although these initialization steps require rather heavy computations, the actual admission control through (19) is simple, because the latter is of the simple form (18). However, note that, since the linearization is optimal only with respect to the chosen n^*, successive connection admissions (and terminations) may move the current traffic mix away from the initial choice n^*, at a vicinity of the domain for which the linearization is overly conservative (see the figure), thus resulting in a waste of network resources. In such a case, a new boundary point close to the current traffic mix should be chosen and the linearization procedure around it should be applied afresh.

We close this subsection by noting that, while the basic asymptotic performance estimate is a standard result in the Theory of Large Deviations (and thus known for many years), its application in the study of broadband networks and, in particular, the results on admission domains and the linearization of their boundaries were originally contributed by Hui [32, 33].

4.3. Asymptotics for large buffers: effective bandwidth theory

We now turn into the study of multiplexers that feature large buffering capabilities. Like previously, we seek to present a theory linking the traffic load and the network resources (viz., the amount of buffer memory and the output link's capacity) to the probability of data loss due to buffer overflow, the latter serving as the performance metric. While in the bufferless setting it was adequate to represent the traffic characteristics through instantaneous rate properties, this subsection deals with large buffers that expose the properties of rate correlations over large time intervals. Therefore, it is necessary to study random variables of the form $V(\tau, \tau + t)$, denoting the amount of data generated over the interval $(\tau, \tau + t]$. It will be assumed throughout that the data process has stationary increments[6], i.e., $V(\tau, \tau + t)$ depends only on the length t of the time-interval, not its origin, and can be denoted simply as $V(t)$. By virtue of stationarity, $\mathbb{E}\,V(t) = \bar{r}t$ for all time-lengths t, $\bar{r}$ being the mean traffic rate. Further stochastic properties of $V(t)$ will be described through the corresponding log-moment generator

$$\phi(\theta, t) \cong \log \mathbb{E}\,e^{\theta V(t)}, \qquad (20)$$

for which two relevant conditions are introduced:

C1 For each θ, the limit $\phi_\infty(\theta) = \lim_{t \to \infty} \frac{\phi(\theta,t)}{t}$ exists and is finite.

C2 $\phi_\infty(\theta)$ is strictly convex and differentiable.

Condition C1 ensures that the traffic is not long-range dependent (a case for which the theory, in the form presented here, does not hold), while Condition C2 is a guarantee that the strict convexity and differentiability of the generator $\phi(\theta, t)$ will also be inherited by the limit.

Under Condition C1, the 'effective bandwidth function' (EBF) of the traffic is defined as

$$a(\theta) = \phi_\infty(\theta)/\theta, \qquad \theta \geq 0. \qquad (21)$$

As a log-moment generator, $\phi(\theta, t)$ is convex in θ, a property also transferred to the limit $\phi_\infty(\theta)$ as well. Thus, according to the discussion early

---

[6] This assumption holds in particular when data are generated according to a stationary rate process $\{r(t),\ t \in \mathbb{R}\}$, since in that case $V(\tau, \tau + t) = \int_\tau^{\tau+t} r(x)\,dx$.

in Subsection 4.2, the EBF $a(\cdot)$ is an increasing function. Furthermore, if Condition C2 also holds, then $a(\cdot)$ is strictly increasing. Lastly, observe that, by virtue of (12),

$$\bar{r} = \lim_{t \to \infty} \frac{E V(t)}{t} = \phi'_\infty(0) = a(0) \leq a(\theta) \leq \lim_{\theta \to \infty} a(\theta) = \lim_{\theta \to \infty} \frac{\phi_\infty(\theta)}{\theta}$$
$$= \lim_{t \to \infty} \frac{\operatorname{ess\ sup} V(t)}{t} \stackrel{\wedge}{=} \hat{r},$$

establishing that the EBF is bounded between mean and peak rate. (The peak rate $\hat{r}$ is with respect to an asymptotically large time-window and may, in some cases, be smaller than the instantaneous peak rate.)

The importance of the EBF is due to the following properties: Assume that traffic of EBF $a(\cdot)$ loads a multiplexer featuring infinite buffer space and an output link of capacity C. Further, assume there is some $\theta > 0$, such that $a(\theta) < C$. Then, it may be proved that the distribution tail of the queue content $Q(t)$ has at all times an exponential upper bound of rate θ. In other words, there exists a constant $d(\theta)$, such that

$$\Pr\{Q(t) > B\} \leq d(\theta) e^{-\theta B}, \qquad \forall t \geq 0, \forall B \geq 0.$$

There is also a "reciprocal" result: If $a(\theta) > C$ the capacity is not large enough and it may be shown that the distribution tail of the queue content cannot be bounded exponentially using rate θ.

These two statements taken together suggest that, in order to achieve an exponential decay of at least rate θ for the overflow probability under increasing buffer size, the system's capacity must be greater than $a(\theta)$. In this case, the achievable decay rate is $\theta^* = \sup\{\theta \mid a(\theta) < C\}$. Obviously, when the EBF is strictly increasing (as when Condition C2 holds), $\theta^* = a^{-1}(C)$. In fact, for this case the following stronger assertion can be made: If, besides Condition C1, C2 also holds, the buffer content $Q(t)$ has a stationary distribution with tail satisfying

$$\lim_{B \to \infty} \frac{-\log \Pr\{Q > B\}}{B} = \theta, \qquad \text{where} \quad \theta = a^{-1}(C). \qquad (22)$$

This result not only establishes asymptotic exponentiality for queue tails, but may also be used to determine the bandwidth requirements, as a function of the buffer size and the QoS level.

Indeed, assume that the multiplexer has a large (but finite) buffer size B and set the requirement that the system overflows with probability at most $e^{-\epsilon}$. (This specifies a quality level equal to ϵ in the terminology of the previous subsection.) Then, by (22), one must ensure that $\theta \geq \epsilon/B$ or, equivalently, $C \geq a(\epsilon/B)$, which is the desired result. Although this last relation is in a form suitable for admission control, it must be remembered that $a(\cdot)$ is the EBF for the *whole* traffic load, thus it depends on the properties of all multiplexed streams. Fortunately, the definition of the EBF by (21) and the additivity of log-moment generators over independent r.vs, ensure that, for a traffic mix $\boldsymbol{n} = (n_1, \ldots, n_k)$, containing n_i streams of class i, for $i = 1, \ldots, k$, the aggregate EBF is simply $a(\theta) = \sum_i n_i a_i(\theta)$. In particular, the relation for the bandwidth requirements becomes

$$a(\epsilon/B) = \sum_{i=1}^{k} n_i a_i(\epsilon/B) \leq C, \qquad (23)$$

specifying a linear boundary of the form (18) for the admission domain and enabling the particularly simple algorithm for incremental admission control discussed in Subsection 4.2.

As a matter of fact, the name 'effective bandwidth' is exactly due to the linearity in (23), as the quantity $a_i(\epsilon/B)$ determines, *independently of the rest of the traffic environment* the amount of bandwidth that must be granted to a source of class i, in order to satisfy the QoS requirements with the given amount of buffering. Due to this independence, each traffic stream behaves, in a sense, like a constant-rate counterpart; for this reason effective bandwidths are sometimes called 'effective rates' or 'equivalent bandwidths'. It is mentioned that originally the term was introduced by [32], in connection with (19). However, since the linearization in (19) is only locally significant, the term is now mostly used in the sense (23), for the large-buffer regime.

Note that the linearity precludes any potential for bandwidth savings due to multiplexing gain. Indeed n traffic streams require bandwidth $C = na(\epsilon/B)$, thus maintaining a constant bandwidth per source C/n, no matter how large n becomes. This is not surprising, as the theory holds asymptotically as the buffer size $B \to \infty$ when the multiplexing gain is negligible, compared to the buffering gain effect.

At this point it is remarked that the effective bandwidth theory was developed through a series of contributions. The asymptotic exponentiality of distribution tails for the stationary queue content and the implications for this on a linear admission domain were originally established for iid, Markovian on/off, and other simple traffic models [23, 25, 34] and were later generalized for the class of arbitrary Markovian fluids [22]. An extended theory that covers more general stationary rate processes followed [36, 11, 24], making explicit use of results from Large Deviations Theory. Furthermore, a modification [21] of the limiting generator $\phi_\infty(\theta)$, using a time scaling more general than linear, allowed the treatment of traffic with long range dependence. See [12] for a review of the effective bandwidth theory along the statistical mechanics viewpoint and [58] for a discussion of resource management techniques based on the effective bandwidth concept. Further references may be found in [35].

Apart from the general properties discussed earlier, the particular form of the EBF $a(\cdot)$ depends on stochastic details specific to the corresponding traffic stream. To review some examples, consider Markovian on/off fluid models, featuring a peak rate r and exponentially distributed on and off sojourns with mean durations τ and σ, respectively. In this case the EBF takes the form

$$a(\theta) = \frac{1}{2}\left(r - (\frac{1}{\tau} + \frac{1}{\sigma})\frac{1}{\theta} + \sqrt{\left(r - (\frac{1}{\tau} + \frac{1}{\sigma})\frac{1}{\theta}\right)^2 + \frac{4r}{\sigma\theta}}\right),$$

a result that originally appeared in [23] and was further exploited in [25]. In the more general case of arbitrary Markovian fluids, traffic is described through a 'rates-matrix' $R = \text{diag}\{r_1, \ldots, r_n\}$ and the infinitesimal generator M of a continuous-time Markov Chain, which governs the transitions between rate values. For this class of models it has been shown [22] that the EBF is $a(\theta) = \lambda_{\max}(R + \frac{1}{\theta}M)$, i.e., the largest eigenvalue of the essentially nonnegative matrix $R + \frac{1}{\theta}M$. A further generalization [39] allows the explicit calculation of effective bandwidths corresponding to semi-Markovian fluids, i.e., models where transitions between rates are still Markovian, but the periods during which rate values are sustained may be arbitrarily distributed (but not heavy-tailed). In this case, the EBF is determined through an implicit function problem, derived from the requirement that the spectral radius of an appropriate nonnegative matrix be equal to unity. For general on/off traffic streams, of peak rate r, this result simplifies as follows: Let $\phi_+(s)$ and $\phi_-(s)$ stand for the log-moment generators corresponding to the distributions of the on and off sojourns, respectively. Then, for any $\theta > 0$, the EBF is $a(\theta) = u(\theta)/\theta$, where $u(\theta)$ is the unique positive solution of

$$\phi_+(r\theta - u) + \phi_-(-u) = 0.$$

We close this subsection by mentioning that, instead of adopting a traffic model and trying to determine the EBF through it (something not always feasible), there are alternative approaches, which target the direct measurement of the EBF, thus bypassing modeling. For work along this line, see, e.g., [19, 13].

4.4. Scaling the system's size

The two previous asymptotic regimes were appropriate for either very large buffers or very small ones. However, there are cases where the buffering resource is neither negligible nor overly dominant and then both the multiplexing- and buffering-gain effects are noticeable and must be taken into account. We now briefly discuss results for this more general setting. The relevant asymptotic regime assumes a large number of traffic streams n and proportionally scaled (large) buffer B and bandwidth C. In other words, $B = bn$ and $C = cn$, maintaining a constant amount of resources per stream, as $n \to \infty$. This type of scaling was originally introduced by [62], in connection with traffic consisting of exponential on/off fluids.

In our setting, each traffic stream is a data generation process, which, as in Subsection 4.3, is assumed to have stationary increments. The generator (20) is again used as the traffic descriptor. (Generalizations, relaxing the assumption on stationarity or the requirement for iid streams exist.) Let the stationary queue content under a load of n traffic streams be denoted as Q_n; then the probability of overflow is $\Pr\{Q_n > bn\}$. The basic result [6, 14] (also [48], for the particular context of general on/off fluids) is that, under some regularity conditions, notably the validity of Condition C1 in Subsection 4.3,

$$\lim_{n \to \infty} \frac{-\log \Pr\{Q_n > bn\}}{n} = I(c, b) \hat{=} \inf_{t>0} \sup_{\theta} \{(ct + b)\theta - \phi(\theta, t)\}. \quad (24)$$

There is also a generalization [18] which, among other things, relaxes the requirement for Condition C1 (by introducing a different time-scaling for the generator) and is appropriate for usage with long range dependent traffic.

For a heuristic explanation of (24) remember that, by Lindley, $Q_n = \sup_{t>0}(V_n(t) - nct)$, where $V_n(t)$ is the total amount of data generated by the n streams over time t. Then (24) is essentially Cramér's asymptotic on $V_n(t) - nct$ (see (15) and (14)), followed by an optimization of the time scale (using Laplace's principle of the 'dominating term'). Note that the appropriate time scale relevant to the result is neither $t = 0$ (as was the case with bufferless systems, where only instantaneous rates were needed) nor $t = \infty$ (which was appropriate for very large buffers), but actually the argument extremizing (24), say t^*. The value of t^* depends on c and b and expresses the relative importance of the multiplexing- and buffering-gain. Indeed, by assuming differentiability, (24) implies that

$$t^* = \frac{\partial I(c,b)}{\partial c} \bigg/ \frac{\partial I(c,b)}{\partial b} = -\frac{\partial b}{\partial c}\bigg|_{I(c,b)=\epsilon}, \quad (25)$$

thus t^* quantifies the (local) tradeoff between bandwidth and buffer for a given quality level per source. More specifically, it is possible to formally define a 'buffer-bandwidth' curve of the form $b(c, \epsilon)$, that describes the amount of buffering required for achieving a desired quality level, given the capacity (all quantities being scaled by the number of sources). It may be shown that this curve is convex [20, 41]. Since, by (25), $dc/db = -1/t^*$, the inverse curve is also convex and decreasing, implying that t^* increases with b. This is intuitive, as larger buffers pronounce more the traffic correlations, requiring a larger time-scale for their representation.

Still in connection with this point, it has been shown [6, 14] that as $b \to \infty$ then $t^* \to \infty$ and $I(c, b)$ tends to the asymptotic (22). Similarly, when $b \to 0$, then also $t^* \to 0$ and $I(c, b)$ tends to the asymptotic of (15) and (14), where in place of the instantaneous rate generator the limit $\lim_{t \to 0} \phi(s/t, t)$ is implied[7]. Thus the theories for small and large buffers may be regarded as special cases of the results in this subsection. Note however, that application of (24) is considerably more difficult than the other asymptotic results, not only because the generator $\phi(\theta, t)$ must be determined for all time-scales instead of at a limiting value, but also because the minimization with respect to time is non-convex (unlike the maximization in θ) and thus difficult to perform numerically.

We close by noting that it is possible to define an admission domain for the more general regime of this subsection. This domain is neither linear (as in Subsection 4.3) nor possessing a convex complement (as in Subsection 4.2). However, it is still possible to obtain a local linearization, around points on the boundary, thus introducing a (locally significant) notion of effective bandwidth for this case too. For more details see [35].

References

[1] D. Anick, D. Mitra, and M. M. Sondhi. "Stochastic theory of a data-handling system with multiple sources". *Bell System Tech. J.*, 61(8):1871–1894, October 1982.

[2] Y. Xiong B. Steyaert and H. Bruneel. An efficient solution technique for discrete-time queues fed by heterogeneous traffic. *International Journal of Communication Systems*, 10:73–86, 1997.

[3] A. Baiocchi, N. Bléfari-Melazzi, A. Roveri, and F. Salvatore. "Stochastic fluid analysis of an ATM multiplexer loaded with heterogeneous on-off sources: an effective computational approach". In *Proc. INFOCOM '92*, pages 3C.3.1–3C.3.10, 1992.

---

[7]This limit coincides with the definition in Subsection 4.2 for fluid traffic.

[4] D. A. Bini, G. Latouche, and B. Meini. Algorithms for tree-like stochastic processes. *In Preparation*, 2002.

[5] C. Blondia. A discrete-time batch markovian arrival process as B-ISDN traffic model. *Belgian Journal of Operations Research, Statistics and Computer Science*, 32(3,4), 1993.

[6] D. D. Botvich and N. G. Duffield. "Large deviations, the shape of the loss curve, and economies of scale in large multiplexers". *Queueing Sys.*, 20:293–320, 1995.

[7] H. Bruneel. Queueing behavior of statistical multiplexers with correlated inputs. *IEEE Transactions on Communications*, COM-36(12):1339–1341, 1988.

[8] H. Bruneel. Performance of discrete-time queueing systems. *Computers and Operations Research*, 20(3):303–320, 1993.

[9] H. Bruneel and B.G. Kim. *Discrete-Time Models for Communication Systems Including ATM*. Kluwer Academic Publishers, Boston, 1993.

[10] J. A. Bucklew. *Large Deviations Techniques in Decision and Estimation*. Wiley, New York, 1990.

[11] C.-S. Chang. "Stability, queue length, and delay of deterministic and stochastic queueing networks". *IEEE Trans. Automat. Control*, 39(5):913–931, May 1994.

[12] C.-S. Chang and J. A. Thomas. "Effective bandwidth in high-speed digital networks". *IEEE JSAC*, 13(6):1091–1100, 1995.

[13] C. Courcoubetis, G. Kesidis, A. Ridder, J. Walrand, and R. Weber. "Admission control and routing in ATM networks using inferences from measured buffer occupancy". *IEEE Trans. Commun.*, 43:1778–1784, 1995.

[14] C. Courcoubetis and R. Weber. "Buffer overflow asymptotics for a switch handling many traffic sources". *J. Appl. Prob.*, 33:886–903, 1996.

[15] B. Steyaert D. Fiems and H. Bruneel. Discrete-time queues with general service times and general server interruptions. *Proceedings of SPIE's International Symposium on Voice, Video and Data Communications (Boston, 6-7 November 2000)*.

[16] A. Dembo and O. Zeitouni. *Large Deviations Techniques and Applications*. Springer, New York, 1998. First edition: Jones & Bartlett, 1993.

[17] J.H. Dshalalow. *Advances in Queueing: theory, methods and open problems*. CRC Press, Boca Raton, 1995.

[18] N. G. Duffield. "Economies of scale in queues with sources having power-law large deviation scalings". *J. Appl. Prob.*, 33:840–857, 1996.

[19] N. G. Duffield, J. T. Lewis, N. O'Connell, R. Russell, and F. Toomey. "Entropy of ATM traffic streams: A tool for estimating QoS parameters". *IEEE JSAC*, 13(6):981–990, 1995.

[20] N. G. Duffield and S. Low. "The cost of quality in networks of aggregate traffic". In *Proc. INFOCOM 1998*, pages 525–532, San Francisco, April 1998.

[21] N. G. Duffield and N. O'Connell. "Large deviations and overflow probabilities for the general single-server queue, with applications". *Math. Proc. Cambridge Philos. Soc.*, 1996.

[22] A. I. Elwalid and D. Mitra. "Effective bandwidth of general Markovian traffic sources and admission control of high speed networks". *IEEE/ACM Trans. Networking*, 1(3):329–343, June 1993.

[23] R. J. Gibbens and P. J. Hunt. "Effective bandwidths for the multi-type UAS channel". *Queueing Sys.*, 9:17–28, 1991.

[24] P. W. Glynn and W. Whitt. "Logarithmic asymptotics for steady-state tail probabilities in a single-server queue". *J. Appl. Prob.*, 31A:131–156, 1994.

[25] R. Guérin, H. Ahmadi, and M. Naghshineh. "Equivalent capacity and its application to bandwidth allocation in high-speed networks". *IEEE JSAC*, 9(7):968–981, September 1991.

[26] E. Desmet H. Bruneel, B. Steyaert and G.H. Petit. Analytic derivation of tail probabilities for queue lengths and waiting times in atm multiserver queues. *European Journal of Operational Research*, 76:563–572, 1994.

[27] Q. He. Queues with marked customers. *Adv. Appl. Prob.*, 28:567–587, 1996.

[28] Q. He. Classification of Markov processes of matrix M/G/1 type with a tree structure and its applications to the MMAP[K]/G[K]/1 queue. *Stochastic Models*, 16(5):407–434, 2000.

[29] Q. He. The versatility of the MMAP[K] and the MMAP[K]/G[K]/1 queue. *Queueing Systems*, 38:397–418, 2001.

[30] Q. He and A.S. Alfa. The discrete time MMAP[K]/PH[K]/1/LCFS-GPR queue and its variants. In *Proc. of the 3rd Int. Conf. on Matrix Analytic Methods*, pages 167–190, Leuven (Belgium), 2000.

[31] Q. He and M.F. Neuts. Markov chains with marked transitions. *Stochastic Processes and their Applications*, 74:37–52, 1998.

[32] J. Hui. "Resource allocation for broadband networks". *IEEE JSAC*, 6(9):1598–1608, December 1988.

[33] J. Hui. *Switching and Traffic Theory for Integrated Broadband Networks*. Kluwer, Boston, 1990.

[34] F. P. Kelly. "Effective bandwidths at multi-class queues". *Queueing Sys.*, 9:5–15, 1991.

[35] F. P. Kelly. "Notes on effective bandwidths". In F. P. Kelly, S. Zachary, and I. Ziedens, editors, *Stochastic Networks. Theory and Applications*, volume 4 of *Royal Statistical Society Lecture Notes Series*, pages 141–168. 1996.

[36] G. Kesidis, J. Walrand, and C.-S. Chang. "Effective bandwidths for multi-class Markov fluids and other ATM sources". *IEEE/ACM Trans. Networking*, 1(4):424–428, August 1993.

[37] K. Kontovasilis and N. Mitrou. "Bursty traffic modeling and efficient analysis algorithms via fluid-flow models for ATM-IBCN". *Ann. Oper. Res.*, 49:279–323, 1994. Special Issue in Methodologies for High Speed Networks.

[38] K. Kontovasilis and N. Mitrou. "Markov modulated traffic with near complete decomposability characteristics and associated fluid queueing models". *Adv. Appl. Prob.*, 27(4):1144–1185, 1995.

[39] K. Kontovasilis and N. Mitrou. "Effective bandwidths for a class of non markovian fluid sources". *Computer Communications Review*, 27(4):263–274, 1997.

[40] L. Kosten. "Stochastic theory of data-handling systems with groups of multiple sources". In H. Rudin and W. Bux, editors, *Performance of Computer Communication Systems*, pages 321–331, Amsterdam, 1984. Elsevier.

[41] K. Kumaran and M. Mandjes. "The buffer-bandwidth trade-off curve is convex". *Queueing Sys.*, 38:471–483, 2001.

[42] D. McDonald and K. Qian. "An approximation method for complete solutions of markov-modulated fluid models". *Queueing Sys.*, 30(3–4):365–384, 1998.

[43] D. Mitra. "Stochastic theory of a fluid model of producers and consumers coupled by a buffer". *Adv. Appl. Prob.*, 20:646–676, 1988.

[44] N. Mitrou, S. Vamvakos, and K. Kontovasilis. "Modelling, parameter assessment and multiplexing analysis of bursty sources with hyperexponentially distributed bursts. *Comput. Networks ISDN Systems*, 27(7):1175–92, 1995.

[45] M.F. Neuts. *Matrix-Geometric Solutions in Stochastic Models, An Algorithmic Approach.* John Hopkins University Press, 1981.

[46] M.F. Neuts. *Structured Stochastic Matrices of M/G/1 type and their applications.* Marcel Dekker, Inc., New York and Basel, 1989.

[47] S. Wittevrongel S. De Vuyst and H. Bruneel. Statistical multiplexing of correlated variable-length packet trains : an analytic performance study. *Journal of the Operational Research Society*, 52(3):318–327, 2001.

[48] A. Simonian and J. Guibert. "Large deviations approximation for fluid queues fed by a large number of on/off sources". *IEEE JSAC*, 13(6):1017–1027, August 1995.

[49] T. E. Stern and A. I. Elwalid. "Analysis of separable Markov-modulated rate models for information-handling systems". *Adv. Appl. Prob.*, 23:105–139, 1991.

[50] B. Steyaert and H. Bruneel. Accurate approximation of the cell loss ratio in atm buffers with multiple servers. In *Performance Modelling and Evaluation of ATM Networks, Volume 1, Chapman and Hall, London, (ISBN: 0-412-71140-0)*, pages 285–296, 1995.

[51] T. Takine, B. Sengupta, and R.W. Yeung. A generalization of the matrix M/G/1 paradigm for Markov chains with a tree structure. *Stochastic Models*, 11(3):411–421, 1995.

[52] B. Van Houdt. *Performance Analysis of Contention Resolution Algorithms in Random Access Systems.* PhD thesis, University of Antwerp (UA), 2001.

[53] B. Van Houdt and C. Blondia. Stability and performance of stack algorithms for random access communication modeled as a tree structured QBD Markov chain. *Stochastic Models*, 17(3):247–270, 2001.

[54] B. Van Houdt and C. Blondia. The delay distribution of a type k customer in a first come first served MMAP[K]/PH[K]/1 queue. *Journal of Applied Probability (to appear)*, 39(1), 2002.

[55] B. Van Houdt and C. Blondia. Robustness of FS-ALOHA. In *Proc of the 4th Int. Conf. on Matrix Analytic Methods (MAM4), to appear,* Adelaide (Australia), July 2002.

[56] B. Van Houdt and C. Blondia. Robustness properties of FS-ALOHA(++): a random access algorithm for dynamic bandwidth allocation. *Journal on Special Topics in Mobile Networking and Applications (MONET) on Performance Evaluation of QoS Architectures in Mobile Networks (submitted)*, 2002.

[57] B. Van Houdt and C. Blondia. Throughput of q-ary splitting algorithms for contention resolution in communication networks. *To appear in Adv. in Performance Analysis*, 2002.

[58] G. de Veciana, G. Kesidis, and J. Walrand. "Resource management in wide-area ATM networks using effective bandwidths". *IEEE JSAC*, 13(6):1081–1090, 1995.

[59] B. Vinck and H. Bruneel. Relationship between delay and buffer contents in atm queues. *Electronics Letters*, 31(12):952–954, 1995.

[60] B. Vinck and H. Bruneel. Delay analysis for single server queues. *Electronics Letters*, 32(9):802–803, 1996.

[61] B. Vinck and H. Bruneel. Delay analysis of multiserver atm buffers. *Electronics Letters*, 32(5):1352–1353, 1996.

[62] A. Weiss. "A new technique for analysing large traffic systems". *Adv. Appl. Prob.*, 18:506–532, 1986.

[63] A. Weiss. "An introduction to large deviations for communication networks". *IEEE JSAC*, 13(6):938–952, August 1995.

[64] S. Wittevrongel and H. Bruneel. Deriving the tail distribution of the buffer contents in a statistical multiplexer with general heterogeneous on/off sources. In *Proceedings of the International Conference on the Performance and Management of Complex Communication Networks, PMCCN '97, (Tsukuba)*, pages 37–56, 1997.

[65] S. Wittevrongel and H. Bruneel. Correlation effects in atm queues due to data format conversions. *Performance Evaluation*, 32(1):35–56, 1998.

[66] S. Wittevrongel and H. Bruneel. Discrete-time atm queues with independent and correlated arrival streams. In *chapter 16 in : Performance Evaluation and Applications of ATM Networks, Kluwer Academic Publishers, Boston*, pages 387–412, 2000.

[67] Y. Xiong and H. Bruneel. Buffer contents and delay for statistical multiplexers with fixed-length packet-train arrival. *Performance Evaluation*, 17(1):31–42, 1993.

[68] Y. Xiong and H. Bruneel. A simple approach to obtain tight upper bounds for the asymptotic queueing behavior of statistical multiplexers with heterogeneous traffic. *Performance Evaluation*, 22(2):159–173, 1995.

[69] R.W. Yeung and A.S. Alfa. The quasi-birth-death type markov chain with a tree structure. *Stochastic Models*, 15(4):639–659, 1999.

[70] R.W. Yeung and B. Sengupta. Matrix product-form solutions for markov chains with a tree structure. *Adv. Appl. Prob.*, 26:965–987, 1994.

ENHANCING WEB PERFORMANCE

Arun Iyengar
IBM T.J. Watson Research Center
aruni@watson.ibm.com

Erich Nahum
IBM T.J. Watson Research Center
nahum@watson.ibm.com

Anees Shaikh
IBM T.J. Watson Research Center
aashaikh@watson.ibm.com

Renu Tewari
IBM Almaden Research Center
tewarir@us.ibm.com

Abstract

This paper provides an overview of techniques for improving Web performance. For improving server performance, multiple Web servers can be used in combination with efficient load balancing techniques. We also discuss how the choice of server architecture affects performance. We examine content distribution networks (CDN's) and the routing techniques that they use. While Web performance can be improved using caching, a key problem with caching is consistency. We present different techniques for achieving varying forms of cache consistency.

Keywords: cache consistency, content distribution networks, Web caching, Web performance, Web servers

Introduction

The World Wide Web has emerged as one of the most significant applications over the past decade. The infrastructure required to support Web traffic is significant, and demands continue to increase at a rapid rate. Highly accessed Web sites may need to serve over a million hits per minute. Additional demands are created by the need to serve dynamic and personalized data.

This paper presents an overview of techniques and components needed to support high volume Web traffic. These include multiple servers at Web sites which can be scaled to accommodate high request rates. Various load balancing techniques have been developed to efficiently route requests to multiple servers. Web sites may also be dispersed or replicated across multiple geographic locations.

Web servers can use several different approaches to handling concurrent requests including processes, threads, event-driven architectures in which a single process is used with non-blocking I/O, and in-kernel servers. Each of these architectural choices has certain advantages and disadvantages. We discuss how these different approaches affect performance.

Over the past few years, a number of content distribution networks (CDN) have been developed to aid Web performance. A CDN is a shared network of servers or caches that deliver content to users on behalf of content providers. The intent of a CDN is to serve content to a client from a CDN server so that response time is decreased over contacting the origin server directly. CDN's also reduce the load on origin servers. This paper examines several issues related to CDN's including their overall architecture and techniques for routing requests. We also provide insight into the performance improvements typically achieved by CDN's.

Caching is a critical technique for improving performance. Caching can take place at several points within the network including clients, servers, and in intermediate proxies between the client and server. A key problem with caching within the Web is maintaining cache consistency. Web objects may have expiration times associated with them indicating when they become obsolete. The problem with expiration times is that it is often not possible to tell in advance when Web data will become obsolete. Expiration times are not sufficient for applications which have strong consistency requirements. Stale cached data and the inability in many cases to cache dynamic and personalized data limits the effectiveness of caching.

The remainder of this paper is organized as follows. Section 1 provides an overview of techniques used for improving performance at a Web site. Section 2 discusses different server architectures. Section 3 presents an overview of content distribution networks. Section 4 discusses Web caching and cache consistency techniques.

1. Improving Performance at a Web Site

Highly accessed Web sites may need to handle peak request rates of over a million hits per minute. Web serving lends itself well to concurrency because transactions from different clients can be handled in parallel. A single Web server can achieve parallelism by multithreading or multitasking between different requests. Additional parallelism and higher throughputs can be achieved by using multiple servers and load balancing requests among the servers.

Figure 1 shows an example of a scalable Web site. Requests are distributed to multiple servers via a load balancer. The Web servers may access one or more databases for creating content. The Web servers would typically contain replicated content so that a request could be directed to any server in the cluster. For storing static files, one way to share them across multiple servers is to use a distributed file system such as AFS or DFS [42]. Copies of files may be cached in one or more servers. This approach works fine if the number of Web servers is not too large and data doesn't change very frequently. For large numbers of servers for which data updates are frequent, distributed file systems can be highly inefficient. Part of the reason for this is the strong consistency model imposed by distributed file systems. Shared file systems require all copies of files to be completely consistent. In order to update a file in one server, all other copies of the file need to be invalidated before the update can take place. These invalidation messages add overhead and latency. At some Web sites, the number of objects updated in temporal proximity to each other can be quite large. During periods of peak updates, the system might fail to perform adequately.

Another method of distributing content which avoids some of the problems of distributed file systems is to propagate updates to servers without requiring the strict consistency guarantees of distributed file systems. Using this approach, updates are propagated to servers without first invalidating all existing copies. This means that at the time an update is made, data may be inconsistent between servers for a little while. For many Web sites, these inconsistencies are not a problem, and the performance benefits from relaxing the consistency requirements can be significant.

1.1. Load Balancing

The load balancer in Figure 1 distributes requests among the servers. One method of load balancing requests to servers is via DNS servers. DNS servers provide clients with the IP address of one of the site's content delivery nodes. When a request is made to a Web site such as http://www.research.ibm.com/compsci/, "www.research.ibm.com" must be translated to an IP address, and DNS servers perform this translation. A name associated with a Web site can map to multiple IP addresses, each associated with a different

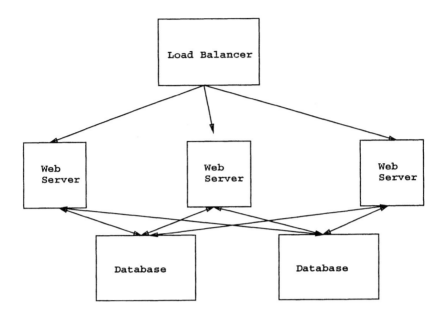

Figure 1. Architecture of a scalable Web site. Requests are directed from the load balancer to one of several Web servers. The Web servers may access one or more databases for creating content.

Web server. DNS servers can select one of these servers using a policy such as round robin [10].

There are other approaches which can be used for DNS load balances which offer some advantages over simple round robin [13]. The DNS server can use information about the number of requests per unit time sent to a Web site as well as geographic information. The Internet2 Distributed Storage Infrastructure Project proposed a DNS that implements address resolution based on network proximity information, such as round-trip delays [8].

One of the problems with load balancing using DNS is that name-to-IP mappings resulting from a DNS lookup may be cached anywhere along the path between a client and a server. This can cause load imbalance because client requests can then bypass the DNS server entirely and go directly to a server [19]. Name-to-IP address mappings have time-to-live attributes (TTL) associated with them which indicate when they are no longer valid. Using small TTL values can limit load imbalances due to caching. The problem with this approach is that it can increase response times [59]. Another problem with this approach is that not all entities caching name-to-IP address mappings obey TTL's which are too short.

Adaptive TTL algorithms have been proposed in which the DNS assigns different TTL values for different clients [12]. A request coming from a client with a high request rate would typically receive a name-to-IP address mapping with a shorter lifetime than that assigned to a client with a low request rate. This prevents a proxy with many clients from directing requests to the same server for too long a period of time.

Another approach to load balancing is using a connection router in front of several back-end servers. Connection routers hide the IP addresses of the back-end servers. That way, IP addresses of individual servers won't be cached, eliminating the problem experienced with DNS load balancing. Connection routing can be used in combination with DNS routing for handling large numbers of requests. A DNS server can route requests to multiple connection routers. The DNS server provides coarse grained load balancing, while the connection routers provide finer grained load balancing. Connection routers also simplify the management of a Web site because back-end servers can be added and removed transparently.

IBM's Network Dispatcher [32] is one example of a connection router which hides the IP address of back-end servers. Network Dispatcher uses Weighted Round Robin for load balancing requests. Using this algorithm, servers are assigned weights. All servers with the same weight receive a new connection before any server with a lesser weight receives a new connection. Servers with higher weights get more connections than those with lower weights, and servers with equal weights get an equal distribution of new connections.

With Network Dispatcher, requests from the back-end servers go directly back to the client. This reduces overhead at the connection router. By contrast, some connection routers function as proxies between the client and server in which all responses from servers go through the connection router to clients.

Network Dispatcher has special features for handling client affinity to selected servers. These features are useful for handling requests encrypted using the Secure Sockets Layer protocol (SSL). When an SSL connection is made, a session key must be negotiated and exchanged. Session keys are expensive to generate. Therefore, they have a lifetime, typically 100 seconds, for which they exist after the initial connection is made. Subsequent SSL requests within the key lifetime reuse the key.

Network dispatcher recognizes SSL requests by the port number (443). It allows certain ports to be designated as "sticky". Network Dispatcher keeps records of old connections on such ports for a designated affinity life span (e.g. 100 seconds for SSL). If a request for a new connection from the same client on the same port arrives before the affinity life span for the previous connection expires, the new connection is sent to the same server that the old connection utilized.

Using this approach, SSL requests from the same client will go to the same server for the lifetime of a session key, obviating the need to negotiate new session keys for each SSL request. This can cause some load imbalance, particularly since the client address seen by Network Dispatcher may actually be a proxy representing several clients and not just the client corresponding to the SSL request. However, the reduction in overhead due to reduced session key generation is usually worth the load imbalance created. This is particularly true for sites which make gratuitous use of SSL. For example, some sites will encrypt all of the image files associated with an HTML page and not just the HTML page itself.

Connection routing is often done at layer 4 of the OSI model in which the connection router does not know the contents of the request. Another approach is to perform routing at layer 7. In layer 7 routing, also known as content-based routing, the router examines requests and makes its routing decisions based on the contents of requests [55]. This allows more sophisticated routing techniques. For example, dynamic requests could be sent to one set of servers, while static requests could be sent to another set. Different quality of service policies could be assigned to different URL's in which the content-based router sends the request to an appropriate server based on the quality of service corresponding to the requested URL. Content-based routing allows the servers at a Web site to be assymetrical. For example, information could be distributed at a Web site so that frequently requested objects are stored on many or all servers, while infrequently requested objects are only stored on a few servers. This reduces the storage overhead of replicating all information on all servers. The content-based router can then use information on how objects are distributed to make correct routing decisions.

The key problem with content-based routing is that the overhead which is incurred can be high [60]. In order to examine the contents of a request, the router must terminate the connection with the client. In a straightforward implementation of content-based routing, the router acts as a proxy between the client and server, and all data exchanged between the client and server go through the router. Better performance is achieved by using a TCP handoff protocol in which the client connection is transferred from the router to a back-end server; this can be done in a manner which is transparent to the client.

A number of client-based techniques have been proposed for load balancing. A few years ago, Netscape implemented a scheme for doing load balancing at the Netscape Web site (before they were purchased by AOL) in which the Netscape browser was configured to pick the appropriate server [49]. When a user accessed the Web site www.netscape.com, the browser would randomly pick a number i between 1 and the number of servers and direct the request to www$i$.netscape.com.

Another client-based technique is to use the client's DNS [23, 58]. When a client wishes to access a URL, it issues a query to its DNS to get the IP address of the site. The Web site's DNS returns a list of IP addresses of the servers instead of a single IP address. The client DNS selects an appropriate server for the client. An alternative strategy is for the client to obtain the list of IP addresses from its DNS and do the selection itself. An advantage to the client making the selection itself is that the client can collect information about the performance of different servers at the site and make an intelligent choice based on this. The disadvantages of client-based techniques is that the Web site loses control over how requests are routed, and such techniques generally require modifications to the client (or at least the client's DNS server).

1.2. Serving Dynamic Web Content

Web servers satisfy two types of requests, static and dynamic. *Static requests* are for files that exist at the time a request is made. *Dynamic requests* are for content that has to be generated by a server program executed at request time. A key difference between satisfying static and dynamic requests is the processing overhead. The overhead of serving static pages is relatively low. A Web server running on a uniprocessor can typically serve several hundred static requests per second. Of course, this number is dependent on the data being served; for large files, the throughput is lower.

The overhead for satisfying a dynamic request may be orders of magnitude more than the overhead for satisfying a static request. Dynamic requests often involve extensive back-end processing. Many Web sites make use of databases, and a dynamic request may invoke several database accesses. These database accesses can consume significant CPU cycles. The back-end software for creating dynamic pages may be complex. While the functionality performed by such software may not appear to be compute-intensive, such middleware systems are often not designed efficiently; commercial products for generating dynamic data can be highly inefficient.

One source of overhead in accessing databases is connecting to the database. Many database systems require a client to first establish a connection with a database before performing a transaction in which the client typically provides authentication information. Establishing a connection is often quite expensive. A naive implementation of a Web site would establish a new connection for each database access. This approach could overload the database with relatively low traffic levels.

A significantly more efficient approach is to maintain one or more long-running processes with open connections to the database. Accesses to the database are then made with one of these long-running processes. That way, multiple accesses to the database can be made over a single connection.

Another source of overhead is the interface for invoking server programs in order to generate dynamic data. The traditional method for invoking server programs for Web requests is via the Common Gateway Interface (CGI). CGI forks off a new process to handle each dynamic request; this incurs significant overhead. There are a number of faster interfaces available for invoking server programs [34]. These faster interfaces use one of two approaches. The first approach is for the Web server to provide an interface to allow a program for generating dynamic data to be invoked as part of the Web server process itself. IBM's GO Web server API (GWAPI) is an example of such an interface. The second approach is to establish long-running processes to which a Web server passes requests. While this approach incurs some interprocess communication overhead, the overhead is considerably less than that incurred by CGI. FastCGI is an example of the second approach [53].

In order to reduce the overhead for generating dynamic data, it is often feasible to generate data corresponding to a dynamic object once, store the object in a cache, and subsequently serve requests to the object from cache instead of invoking the server program again [33]. Using this approach, dynamic data can be served at about the same rate as static data.

However, there are types of dynamic data that cannot be precomputed and served from a cache. For instance, dynamic requests that cause a side effect at the server such as a database update cannot be satisfied merely by returning a cached page. As an example, consider a Web site that allows clients to purchase items using credit cards. At the point at which a client commits to buying something, that information has to be recorded at the Web site; the request cannot be solely serviced from a cache.

Personalized Web pages can also negatively affect the cacheability of dynamic pages. A personalized Web page contains content specific to a client, such as the client's name. Such a Web page could not be used for another client. Therefore, caching the page is of limited utility since only a single client can use it. Each client would need a different version of the page.

One method which can reduce the overhead for generating dynamic pages and enable caching of some parts of personalized pages is to define these pages as being composed of multiple fragments [15]. In this approach, a complex Web page is constructed from several simpler fragments. A fragment may recursively embed other fragments. This is efficient because the overhead for assembling a Web page from simpler fragments is usually minor compared to the overhead for constructing the page from scratch, which can be quite high.

The fragment-based approach also makes it easier to design Web sites. Common information that needs to be included on multiple Web pages can be created as a fragment. In order to change the information on all pages, only the fragment needs to be changed.

In order to use fragments to allow partial caching of personalized pages, the personalized information on a Web page is encapsulated by one or more fragments that are not cacheable, but the other fragments in the page are. When serving a request, a cache composes pages from its constituent fragments, many of which are locally available. Only personalized fragments have to be created by the server. As personalized fragments typically constitute a small fraction of the entire page, generating only them would require lower overhead than generating all of the fragments in the page.

Generating Web pages from fragments provides other benefits as well. Fragments can be constructed to represent entities that have similar lifetimes. When a particular fragment changes but the rest of the Web page stays the same, only the fragment needs to be invalidated or updated in the cache, not the entire page. Fragments can also reduce the amount of cache space taken up by multiple pages with common content. Suppose that a particular fragment is contained in 2000 popular Web pages which should be cached. Using the conventional approach, the cache would contain a separate version of the fragment for each page resulting in as many as 2000 copies. By contrast, if the fragment-based method of page composition is used, only a single copy of the fragment needs to be maintained.

A key problem with caching dynamic content is maintaining consistent caches. It is advantageous for the cache to provide a mechanism, such as an API, allowing the server to explicitly invalidate or update cached objects so that they don't become obsolete. Web objects may be assigned expiration times that indicate when they should be considered obsolete. Such expiration times are generally not sufficient for allowing dynamic data to be cached properly because it is often not possible to predict accurately when a dynamic page will change.

2. Server Performance Issues

A central component of the response time seen by Web users is, of course, the performance of the origin server that provides the content. There is great interest, then, understanding the performance of Web servers: How quickly can they respond to requests? How well do they scale with load? Are they capable of operating under overload, i.e., can they maintain some level of service even when the requested load far outstrips the capacity of the server?

A Web server is an unusual piece of software in that it must communicate with potentially thousands of remote clients simultaneously. The server thus must be able to deal with a large degree of *concurrency*. A server cannot simply respond to each client in a non-preemptive, first-come first-serve manner, for several reasons. Clients are typically located far away over the wide-area Internet, and thus connection lifetimes can last many seconds or even minutes. Particularly with HTTP 1.1, a client connection may be open but idle for some time before a new request is submitted. Thus a server can have many concurrent connections open, and should be able do work for one connection when another is quiescent. Another reason is that a client may request a file which is not resident in memory. While the server CPU waits for the disk to retrieve the file, it can work on responding to another client. For these and other reasons, a server must be able to multiplex the work it has to do through some form of concurrency.

A fundamental factor which affects the performance of a Web server is the *architectural model* that it uses to implement that concurrency. Generally, Web servers can be implemented using one of four architectures: processes, threads, event-driven, and in-kernel. Each approach has its advantages and disadvantages which we go into more detail below. A central issue in this decision of which model to use is what sort of performance optimizations are available under that model. Another is how well that model *scales* with the workload, i.e., how efficiently it can handle growing numbers of clients.

2.1. Process-Based Servers

Processes are perhaps the most common form of providing concurrency. The original NCSA server and the widely-known Apache server [2] use processes as the mechanism to handle large numbers of connections. In this model, a process is created for each new request, which can block when necessary, for example waiting for data to become available on a socket or for file I/O to be available from the disk. The server handles concurrency by creating multiple processes.

Processes have two main advantages. First, they are consistent with a programmers' way of thinking, allowing the developer to proceed in a step-by-step fashion without worrying about managing concurrency. Second, they provide isolation and protection between different clients. If one process hangs or crashes, the other processes should be unaffected.

The main drawback to processes is performance. Processes are relatively heavyweight abstractions in most operating systems, and thus creating them, deleting them, and switching context between them is expensive. Apache, for example, tries to ameliorate these costs by pre-forking a number of processes and only destroys them if the load falls below a certain threshold. However, the costs are still significant, as each process requires memory to be allocated to them. As the number of processes grow, large amounts of memory are used which puts pressure on the virtual memory system, which could use the memory for other purposes, such as caching frequently-accessed data. In addition, sharing information, such as a cached file, across processes can be difficult.

2.2. Thread-Based Servers

Threads are the next most common form of concurrency. Servers that use threads include JAWS [31] and Sun's Java Web Server [64]. Threads are similar to processes but are considered lighter-weight. Unlike processes, threads share the same address space and typically only provide a separate stack for each thread. Thus, creation costs and context-switching costs are usually much lower than for processes. In addition, sharing between threads is much easier. Threads also maintain the abstraction of an isolated environment much like processes, although the analogy is not exact since programmers must worry more about issues like synchronization and locking to protect shared data structures.

Threads have several disadvantages as well. Since the address space is shared, threads are not protected from one another the way processes are. Thus, a poorly programmed thread can crash the whole server. Threads also require proper operating system support, otherwise when a thread blocks on something like a file I/O, the whole address space will be stopped.

2.3. Event-Driven Servers

The third form of concurrency is known as the *event-driven* architecture. Servers that use this method include Flash [56] and Zeus [72]. With this architecture, a single process is used with *non-blocking* I/O. Non-blocking I/O is a way of doing asynchronous reads and writes on a socket or file descriptor. For example, instead of a process reading a file descriptor and blocking until data is available, an event-driven server will return immediately if there is no data. In turn, the O.S. will let the server process know when a socket or file descriptor is ready for reading or writing through a *notification mechanism*. This notification mechanism can be an active one such as a signal handler, or a passive one requiring the process to ask the O.S. such as the `select()` system call. Through these mechanisms the server process will essentially respond to events and is typically guaranteed to never block.

Event-driven servers have several advantages. First, they are very fast. Zeus is frequently used by hardware vendors to generate high Web server numbers with the SPECWeb99 benchmark [61]. Sharing is inherent, since there is only one process, and no locking or synchronization is needed. There are no context-switch costs or extra memory consumption that are the case with threads or processes. Maximizing concurrency is thus much easier than with the previous approaches.

Event-driven servers have downsides as well. Like threads, a failure can halt the whole server. Event-driven servers can tax operating system resource limits, such as the number of open file descriptors. Different operating systems have varying levels of support for asynchronous I/O, so a fully event-driven server may not be possible on a particular platform. Finally, event-driven servers require a different way of thinking from the programmer, who must understand and account for the ways in which multiple requests can be in varying stages of progress simultaneously. In this approach, the degree of concurrency is fully exposed to the developer, with all the attendant advantages and disadvantages.

2.4. In-Kernel Servers

The fourth and final form of server architectures is the *in-kernel* approach. Servers that use this method include AFPA [36] and Tux [66]. All of the previous architectures place the Web server software in user space; in this approach the HTTP server is in kernel space, tightly integrated with the host TCP/IP stack.

The in-kernel architecture has the advantages that it is extremely fast, since potentially expensive transitions to user space are completely avoided. Similarly, no data needs to be copied across the user-kernel boundary, another costly operation.

The disadvantages for in-kernel approaches are several. First, it is less robust to programming errors; a server fault can crash the whole machine, not just the server! Development is much harder, since kernel programming is more difficult and much less portable than programming user-space applications. Kernel internals of Linux, FreeBSD, and Windows vary considerably, making deployment across platforms more work. The socket and thread APIs, on the other hand, are relatively stable and portable across operating systems.

Dynamic content poses an even greater challenge for in-kernel servers, since an arbitrary program may be invoked in response to a request for dynamic content. A full-featured in-kernel web server would need to have a PHP engine or Java runtime interpreter loaded in with the kernel! The way current in-kernel servers deal with this issue is to restrict their activities to the static content component of Web serving, and pass dynamic content requests to a complete server in user space, such as Apache. For example, many entries in the SPECWeb99 site [61] that use the Linux operating system use this hybrid approach, with Tux serving static content in the kernel and Apache handling dynamic requests in user space.

2.5. Server Performance Comparison

Since we are concerned with performance, it is thus interesting to see how well the different server architectures perform. To evaluate them, we took a experimental testbed setup and evaluate the performance using a synthetic workload generator [51] to saturate the servers with requests for a range of web documents. The clients were eight 500 MHz PC's running FreeBSD, and the server was a 400 MHz PC running Linux 2.4.16. Each client had a 100 mbps Ethernet connected to a gigabit switch, and the server was connected to the switch using Gigabit Ethernet. Three servers were evaluated as representatives

Enhancing Web Performance

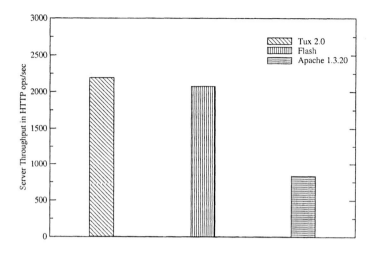

Figure 2. Server Throughput

of their architecture: Apache as a process-based server, Flash as an event-driven server, and Tux as an in-kernel server.

Figure 2 shows the server throughput in HTTP operations/sec of the three servers. As can be seen, Tux, the in-kernel server, is the fastest at 2193 ops/sec. However, Flash is only 10 percent slower at 2075 ops/sec, despite being implemented in user space. Apache, on the other hand, is significantly slower at 875 ops/sec. Figure 3 shows the server response time for the three servers. Again, Tux is the fastest, at 3 msec, Flash second at 5 msec, and Apache slowest at 10 msec.

Since multiple examples of each type of server architecture exist, there is clearly no consensus for what is the best model. Instead, it may be that different approaches are better suited for different scenarios. For example, the in-kernel approach may be most appropriate for dedicated server appliances, or as CDN nodes, whereas a back-end dynamic content server will rely on the full generality of a process-based server like Apache. Still, web site operators should be aware of how the choice of architecture will affect Web server performance.

3. CDNs: Improved Web Performance through Distribution

End-to-end Web performance is influenced by numerous factors such as client and server network connectivity, network loss and delay, server load, HTTP protocol version, and name resolution delays. The content-serving architecture has a significant impact on some of these factors, as well factors

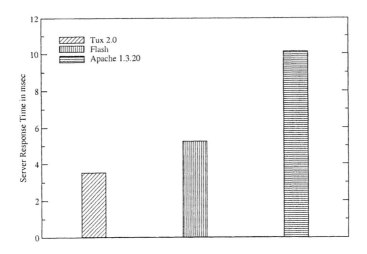

Figure 3. Server Response Time

not related to performance such as cost, reliability, and ease of management. In a traditional content-serving architecture all clients request content from a single location, as shown in Figure 4. In this architecture, scalability and performance are improved by adding servers, without the ability to address poor performance due to problems in the network. Moreover, this approach can be expensive since the site must be overprovisioned to handle unexpected surges in demand.

One way to address poor performance due to network congestion, or flash crowds at servers, is to distribute content to servers or caches located closer to the edges of the network, as shown in Figure 5. Such a distributed network of servers comprises a content distribution network (CDN). A CDN is simply a network of servers or caches that deliver content to users on behalf of content providers. The intent of a CDN is to serve content to a client from a CDN server such that the response-time performance is improved over contacting the origin server directly. CDN servers are typically shared, delivering content belonging to multiple Web sites though all servers may not be used for all sites.

CDNs have several advantages over traditional centralized content-serving architectures, including [67]:

- improving client-perceived response time by bringing content closer to the network edge, and thus closer to end-users

- off-loading work from origin servers by serving larger objects, such as images and multimedia, from multiple CDN servers

Enhancing Web Performance

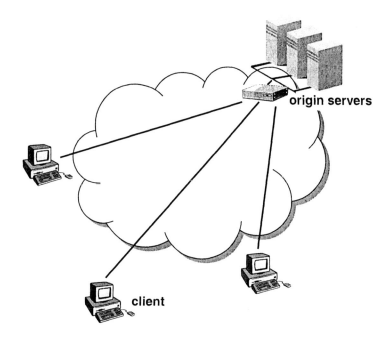

Figure 4. Traditional centralized content-serving architecture

- reducing content provider costs by reducing the need to invest in more powerful servers or more bandwidth as user population increases

- improving site availability by replicating content in many distributed locations

CDN servers may be configured in tree-like hierarchies [71] or clusters of cooperating proxies that employ content-based routing to exchange data [28]. Commercial CDNs also vary significantly in their size and service offerings. CDN deployments range from a few tens of servers (or server clusters), to over ten thousand servers placed in hundreds of ISP networks. A large footprint allows a CDN service provider (CDSP) to reach the majority of clients with very low latency and path length.

Content providers use CDNs primarily for serving static content like images or large stored multimedia objects (e.g., movie trailers and audio clips). A recent study of CDN-served content found that 96% of the objects served were images [41]. However, the remaining few objects accounted for 40–60% of the bytes served, indicating a small number of very large objects. Increasingly, CDSPs offer services to deliver streaming media and dynamic data such as localized content or targeted advertising.

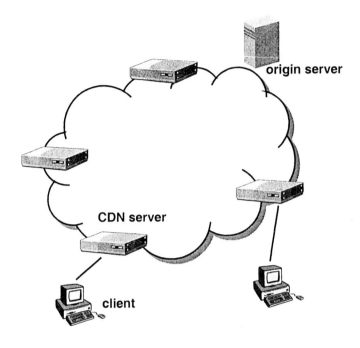

Figure 5. Distributed CDN architecture

3.1. CDN Architectural Elements

As illustrated in Figure 6, CDNs have three key architectural elements in addition to the CDN servers themselves: a distribution system, an accounting/billing system, and a request-routing system [18]. The distribution system is responsible for moving content from origin servers into CDN servers and ensuring data consistency. Section 4.4 describes some techniques used to maintain consistency in CDNs. The accounting/billing system collects logs of client accesses and keeps tracks CDN server usage for use primarily in administrative tasks. Finally, the request-routing system is responsible for directing client requests to appropriate CDN servers. It may also interact with the distribution system to keep an up-to-date view of which content resides on which CDN servers.

The request-routing system operates as shown in Figure 7. Clients access content from the CDN servers by first contacting a request router (step 1). The request router makes a server selection decision and returns a server assignment to the client (step 2). Finally, the client retrieves content from the specified CDN server (step 3).

Enhancing Web Performance

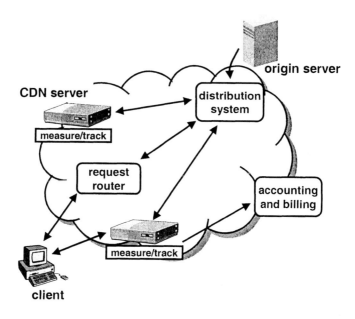

Figure 6. CDN architectural elements

3.2. CDN Request-Routing

Clearly, the request-routing system has a direct impact on the performance of the CDN. A poor server selection decision can defeat the purpose of the CDN, namely to improve client response time over accessing the origin server. Thus, CDNs typically rely on a combination of static and dynamic information when choosing the best server. Several criteria are used in the request-routing decision, including the content being requested, CDN server and network conditions, and client proximity to the candidate servers.

The most obvious request routing strategy is to direct the client to a CDN server that hosts the content being requested. This is complicated, however, if the request router does not know the content being requested, for example if request-routing is done in the context of name resolution. In this case the request contains only a server name (e.g., www.service.com) as opposed to the full HTTP URL.

For good performance the client should be directed to a relatively unloaded CDN server. This requires that the request router actively monitor the state of CDN servers. If each CDN location consists of a cluster of servers and local load-balancer, it may be possible to query a server-side agent for server load information, as shown in Figure 8. After the client makes its request, the

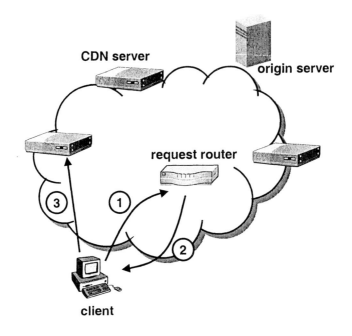

Figure 7. CDN request-routing

request router consults an agent at each CDN site load-balancer (step 2), and returns an appropriate answer back to the client.

As Web response time is heavily influenced by network conditions, it is important to choose a CDN server to which the client has good connectivity. Upon receiving a client request, the request router can ask candidate CDN servers to measure network latency to the client using ICMP echo (i.e., ping) and report the measured values. The request router then responds to the client request with the CDN server reporting the lowest delay. Since these measurements are done on-line, this technique has the advantage of adapting the request-routing decision to the most current network network. On the other hand, it introduces additional latency for the client as the request router waits for responses from the CDN servers.

A common strategy used in CDN request-routing is to choose a server "near" the client, where proximity is defined in terms of network topology, geographic distance, or network latency. Examples of proximity metrics include autonomous system (AS) hops or network hops. These metrics are relatively static compared with server load or network performance, and are also easier to measure.

Enhancing Web Performance

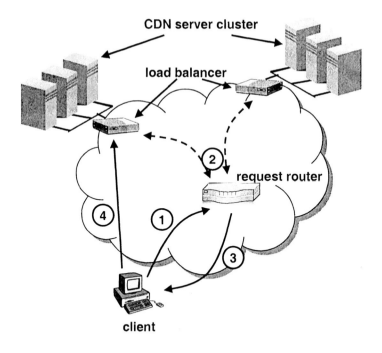

Figure 8. Interaction between request router and CDN servers

Note that it is unlikely that any one of these metrics will be suitable in all cases. Most request routers use a combination of proximity and network or server load to make server selection decisions. For example, client proximity metrics can be used to assign a client to a "default" CDN server, which provides good performance most of the time. The selection can be temporarily changed if load monitoring indicates that the default server is overloaded.

Request-routing techniques fall into three main categories: transport-layer mechanisms, application-layer redirection, and DNS-based approaches [6]. Transport-layer request routers use information in the transport-layer headers to determine which CDN server should serve the client. For example, the request router can examine the client IP address and port number in a TCP SYN packet and forward the packet to an appropriate CDN server. The target CDN server establishes the TCP connection and proceeds to serve the requested content. Forward traffic (including TCP acknowledgements) from the client to the target server continues to be sent to the request router and forwarded to the CDN server. The bulk of traffic (i.e., the requested content) will travel on the direct path from the CDN server to the client.

Application-layer request-routing has access to much more information about the content being requested. For example, the request-router can use HTTP headers like the URL, HTTP cookies, and Language. A simple implementation of an application-layer request router is a Web server that receives client requests and returns an HTTP redirect (e.g., return code 302) to the client indicating the appropriate CDN server. The flexibility afforded by this approach comes at the expense of added latency and overhead, however, since it requires TCP connection establishment and HTTP header parsing.

With request-routing based on the Domain Name System (DNS), clients are directed to the nearest CDN server during the name resolution phase of Web access. Typically, the authoritative DNS server for the domain or subdomain is controlled by the CDSP. In this scheme, a specialized DNS server receives name resolution requests, determines the location of the client and returns the address of a nearby CDN server or a referral to another nameserver. The answer may only be cached at the client-side for a short time so that the request router can adapt quickly to changes in network or server load. This is achieved by setting the associated time-to-live (TTL) field in the answer to a very small value (e.g., 20 seconds).

DNS-based request routing may be implemented with either full- or partial-site content delivery [41]. In full-site delivery, the content provider delegates authority for its domain to the CDSP or modifies its own DNS servers to return a referral (CNAME record) to the CDSPs DNS servers. In this way, all requests for www.company.com, for example, are resolved to a CDN server which then delivers all of the content. With partial-site delivery, the content provider modifies its content so that links to specific objects have hostnames in a domain for which the CDSP is authoritative. For example, links to http://www.company.com/image.gif are changed to http://cdsp.net/company.com/image.gif. In this way, the client retrieves the base HTML page from the origin server but retrieves embedded images from CDN servers to improve performance.

The appeal of DNS-based server selection lies in both its simplicity – it requires no change to existing protocols, and its generality – it works across any IP-based application regardless of the transport-layer protocol being used. This has led to adoption of DNS-based request routing as the *de facto* standard method by many CDSPs and equipment vendors. Using the DNS for request-routing does have some fundamental drawbacks, however, some of which have been recently studied and evaluated [59, 45, 6].

3.3. CDN Performance Studies

Several research studies have recently tried to quantify the extent to which CDNs are able to improve response-time performance. An early study by Johnson *et al.* focused on the quality of the request-routing decision [35]. The study compared two CDSPs that use DNS-based request-routing. The methodology was to measure the response time to download a single object from the CDN server assigned by the request router and the time to download it from all other CDN servers that could be identified. The findings suggested that the server selection did not always choose the best CDN server, but it was effective in avoiding poorly performing servers, and certainly better than choosing a CDN server randomly. The scope of the study was limited, however, since only three client locations were considered, performance was compared for downloading only one small object, and there was no comparison with downloading from the origin server.

A study done in the context of developing the request mirroring Medusa Web proxy, evaluated the performance of one CDN (Akamai) by downloading the same objects from CDN servers and origin servers [37]. The study was done only for a single-user workload, but showed significant performance improvement for those objects that were served by the CDN, when compared with the origin server.

More recently, Krishnamurthy *et al.* studied the performance of a number of commercial CDNs from the vantage point of approximately 20 clients [41]. The authors conclude that CDN servers generally offer much better performance than origin servers, though the gains were dependent on the level of caching and the HTTP protocol options. There were also significant differences in download times from different CDNs. The study finds that, for some CDNs, DNS-based request routing significantly hampers performance due to multiple name lookups.

4. Cache Consistency

Caching has proven to be an effective and practical solution for improving the scalability and performance of Web servers. Static Web page caching has been applied both at browsers at the client, or at intermediaries that include isolated proxy caches or multiple caches or servers within a CDN network. As with caching in any system, maintaining cache consistency is one of the main issues that a Web caching architecture needs to address. As more of the data on the Web is dynamically assembled, personalized, and constantly changing, the challenges of efficient consistency management become more pronounced. To prevent stale information from being transmitted to clients, an intermediary cache must ensure that the locally cached data is consistent with that stored on servers. The exact cache consistency mechanism and the degree of consistency employed by an intermediary depends on the nature of the cached data; not all types of data need the same level of consistency guarantees. Consider the following example.

Example 1 Online auctions: *Consider a Web server that offers online auctions over the Internet. For each item being sold, the server maintains in-*

formation such as its latest bid price (which changes every few minutes) as well as other information such as photographs and reviews for the item (all of which change less frequently). Consider an intermediary that caches this information. Clearly, the bid price returned by the intermediary cache should always be consistent with that at the server. In contrast, reviews of items need not always be up-to-date, since a user may be willing to receive slightly stale information.

The above example shows that an intermediary cache will need to provide different degrees of consistency for different types of data. The degree of consistency selected also determines the mechanisms used to maintain it, and the overheads incurred by both the server and the intermediary.

4.1. Degrees of Consistency

In general the degrees of consistency that an intermediary cache can support fall into the following four categories.

- *strong consistency*: A cache consistency level that always returns the results of the latest (committed) write at the server is said to be strongly consistent. Due to the unbounded message delays in the Internet, no cache consistency mechanism can be strongly consistent in this idealized sense. Strong consistency is typically implemented using a two-phase message exchange along with timeouts to handle unbounded delays.

- *delta consistency*: A consistency level that returns data that is never outdated by more than δ time units, where δ is a configurable parameter, with the last committed write at the server is said to be delta consistent. In practice the value of delta should be larger than t which is the network delay between the server and the intermediary at that instant, i.e., $t < \delta \leq \infty$.

- *weak consistency*: For this level of consistency, a read at the intermediary does not necessarily reflect the last committed write at the server but some correct previous value.

- *mutual consistency*: A consistency guarantee in which a group of objects are mutually consistent with respect to each other. In this case some objects in the group cannot be more current than the others. Mutual consistency can co-exist with the other levels of consistency.

Strong consistency is useful for mirror sites that need to reflect the current state at the server. Some applications based on financial transactions may also require strong consistency. Certain types of applications can tolerate stale data as long as it is within some known time bound. For such applications delta consistency is recommended. Delta consistency assumes that there is a bounded

Enhancing Web Performance 117

Overheads	Polling	Periodic polling	Invalidates	Leases	TTL
File Transfer	W'	$W' - \delta$	W'	W'	W'
Control Msgs.	2R-W'	$2R/t - (W' - \delta)$	2W'	2W'	W'
Staleness	0	t	0	0	0
Write delay	0	0	notify(all)	$min(t, notify(all_t))$	0
Server State	None	None	All	All_t	None

Table 1. Overheads of Different Consistency Mechanisms. Key: t is the period in periodic polling or the lease duration in the leases approach. W' is the number of non-consecutive writes. All consecutive writes with no interleaving reads is counted as a single write. R is the number of reads. δ is the number of writes that were not notified to the intermediary as only weak consistency was provided.

communication delay between the server and the intermediary cache. Mutual consistency is useful when a certain set of objects at the intermediary (e.g., the fragments within a sports score page, or within a financial page) need to be consistent with respect to each other. To maintain mutual consistency the objects need to be atomically invalidated such that they all either reflect the new version or maintain the earlier stale version.

Most intermediaries deployed in the Internet today provide only weak consistency guarantees [29, 62]. Until recently, most objects stored on Web servers were relatively static and changed infrequently. Moreover, this data was accessed primarily by humans using browsers. Since humans can tolerate receiving stale data (and manually correct it using browser reloads), weak cache consistency mechanisms were adequate for this purpose. In contrast, many objects stored on Web servers today change frequently and some objects (such as news stories or stock quotes) are updated every few minutes [7]. Moreover, the Web is rapidly evolving from a predominantly read-only information system to a system where collaborative applications and program-driven agents frequently read as well as write data. Such applications are less tolerant of stale data than humans accessing information using browsers. These trends argue for augmenting the weak consistency mechanisms employed by today's proxies with those that provide strong consistency guarantees in order to make caching more effective. In the absence of such strong consistency guarantees, servers resort to marking data as uncacheable, and thereby reduce the effectiveness of proxy caching.

4.2. Consistency Mechanisms

The mechanisms used by an intermediary and the server to provide the degrees of consistency described earlier fall into 3 categories: i) *client-driven*, ii) *server-driven*, and iii) *explicit* mechanisms .

Server-driven mechanisms, referred to as *server-based invalidation*, can be used to provide strong or delta consistency guarantees [69]. Server-based invalidation, requires the server to notify proxies when the data changes. This approach substantially reduces the number of control messages exchanged between the server and the intermediary (since messages are sent only when an object is modified). However, it requires the server to maintain per-object state consisting of a list of all proxies that cache the object; the amount of state maintained can be significant especially at popular Web servers. Moreover, when an intermediary is unreachable due to network failures, the server must either delay write requests until it receives all the acknowledgments or a timeout occurs, or risk violating consistency guarantees. Several new protocols have been proposed recently to provide delta and strong consistency using server-based invalidations. Web cache invalidation protocol (WCIP) is one such proposal for propagating server invalidations using application-level multicast while providing delta consistency [43]. Web content distribution protocol (WCDP) is another proposal that supports multiple consistency levels using a request-response protocol that can be scaled to support distribution hierarchies [65].

The client-driven approach, also referred to as *client polling*, requires that intermediaries poll the server on *every read* to determine if the data has changed [69]. Frequent polling imposes a large message overhead and also increases the response time (since the intermediary must await the result of its poll before responding to a read request). The advantage, though, is that it does not require any state to be maintained at the server, nor does the server ever need to delay write requests (since the onus of maintaining consistency is on the intermediary).

Most existing proxies provide only weak consistency by (i) explicitly providing a server specified lifetime of an object (referred to as the *time-to-live (TTL)* value), or (ii) by *periodic polling* of the the server to verify that the cached data is not stale [14, 29, 62]. The TTL value is sent as part of the HTTP response in an Expires tag or using the Cache-Control headers. However, *a priori* knowledge of when an object will be modified is difficult in practice and the degree of consistency is dependent on the clock skew between the server and the intermediaries. With periodic polling the length of the period determines the extent of the object staleness. In either case, modifications to the object before its TTL expires or between two successive polls causes the intermediary to return stale data. Thus both mechanisms are heuristics and provide only weak consistency guarantees. Hybrid approaches where the server specifies a time-to-live value for each object and the intermediary polls the server only when the TTL expires also suffer from these drawbacks.

Server-based invalidation and client polling form two ends of a spectrum. Whereas the former minimizes the number of control messages exchanged but

Enhancing Web Performance

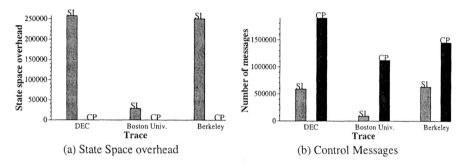

Figure 9. Efficacy of server-based invalidation and client polling for three different trace workloads (DEC, Berkeley, Boston University). The figure shows that server-based invalidation has the largest state space overhead; client polling has the highest control message overhead

may require a significant amount of state to be maintained, the latter is stateless but can impose a large control message overhead. Figure 9 quantitatively compares these two approaches with respect to (i) the server overhead, (ii) the network overhead, and (iii) the client response time. Due to their large overheads, neither approach is appealing for Web environments. A strong consistency mechanism suitable for the Web must not only reduce client response time, but also balance both network and server overheads.

One approach that provides strong consistency, while providing a smooth tradeoff between the state space overhead and the number of control messages exchanged, is *leases* [27]. In this approach, the server grants a lease to each request from an intermediary. The lease duration denotes the interval of time during which the server agrees to notify the intermediary if the object is modified. After the expiration of the lease, the intermediary must send a message requesting renewal of the lease. The duration of the lease determines the server and network overhead. A smaller lease duration reduces the server state space overhead, but increases the number of control (lease renewal) messages exchanged and vice versa. In fact, an infinite lease duration reduces the approach to server-based invalidation, whereas a zero lease duration reduces it to client-polling. Thus, the leases approach spans the entire spectrum between the two extremes of server-based invalidation and client-polling.

The concept of a lease was first proposed in the context of cache consistency in distributed file systems [27]. Recently some research groups have begun investigating the use of leases for maintaining consistency in Web intermediary caches. The use of leases for Web proxy caches was first alluded to in [11] and was subsequently investigated in detail in [69]. The latter effort focused on the design of *volume leases* – leases granted to a collection of objects – so as to reduce (i) the lease renewal overhead and (ii) the blocking overhead at the server due to unreachable proxies. Other efforts have focused on extending leases to hierarchical proxy cache architectures [70, 71]. The adaptive leases effort described analytical and quantitative results on how to select the optimal lease duration based on the server and message exchange overheads [21].

A qualitative comparison of the overheads of the different consistency mechanisms is shown in Table 1. The message overheads of an invalidation-based or lease-based approach is smaller than that of polling especially when reads dominate writes, as in the Web environment.

4.3. Invalidates and Updates

With server-driven consistency mechanisms, when an object is modified, the origin server notifies each "subscribing" intermediary. The notification consists of either an invalidate message or an updated (new) version of the object. Sending an invalidate message causes an intermediary to mark the object as invalid; a subsequent request requires the intermediary to fetch the object from the server (or from a designated site). Thus, each request after a cache invalidate incurs an additional delay due to this remote fetch. An invalidation adds to 2 control messages and a data transfer (an invalidation message, a read request on a miss, and a new data transfer) along with the extra latency. No such delay is incurred if the server sends out the new version of the object upon modification. In an update-based scenario, subsequent requests can be serviced using locally cached data. A drawback, however, is that sending updates incurs a larger network overhead (especially for large objects). This extra effort is wasted if the object is never subsequently requested at the intermediary. Consequently, cache invalidates are better suited for less popular objects, while updates can yield better performance for frequently requested small objects. Delta encoding techniques have been designed to reduce the size of the data transferred in an update by sending only the changes to the object[40]. Note that delta encoding is not related to delta consistency. Updates, however, require better security guarantees and make strong consistency management more complex. Nevertheless, updates are useful for mirror sites where data needs to be "pushed" to the replicas when it changes. Updates are also useful for pre-loading caches with content that is expected to become popular in the near future.

A server can dynamically decide between invalidates and updates based on the characteristics of an object. One policy could be to send updates for objects whose popularity exceeds a threshold and to send invalidates for all other objects. A more complex policy is to take both popularity and object size into account. Since large objects impose a larger network transfer overhead, the server can use progressively larger thresholds for such objects (the larger an object, the more popular it needs to be before the server starts sending updates).

The choice between invalidation and updates also affects the implementation of a strong consistency mechanism. For invalidations only, with a strong consistency guarantee, the server needs to wait for all acknowledgments of the invalidation message (or a timeout) to commit the write at the server. With updates, on the other hand, the server updates are not immediately committed at the intermediary. Only after the server receives all the acknowledgments (or a timeout) and then sends a commit message to all the intermediaries is the new update version committed at the intermediary. Such two-phase message exchanges are expensive in practice and are not required for weaker consistency guarantees.

4.4. Consistency Management for CDNs

An important issue that must be addressed in a CDN is that of *consistency maintenance*. The problem of consistency maintenance in the context of a single proxy used several techniques such as time-to-live (TTL) values, client-polling, server-based invalidation, adaptive refresh [63], and leases [68]. In the simplest case, a CDN can employ these techniques at each individual CDN server or proxy – each proxy assumes responsibility for maintaining consistency of data stored in its cache and interacts with the server to do so independently of other proxies in the CDN. Since a typical CDN may consist of hundreds or thousands of proxies (e.g., Akamai currently has a footprint of more than 14,000 servers), requiring each proxy to maintain consistency independently of other proxies is not scalable from the perspective of the origin servers (since the server will need to individually interact with a large number of proxies). Further, consistency mechanisms designed from the perspective of a single proxy (or a small group of proxies) do not scale well to large CDNs. The leases approach, for instance, requires the origin server to maintain per-proxy state for each cached object. This state space can become excessive if proxies cache a large number of objects or some objects are cached by a large number of proxies within a CDN.

A cache consistency mechanism for hierarchical proxy caches was discussed in [71]. The approach does not propose a new consistency mechanism, rather it examines issues in instantiating existing approaches into a hierarchical proxy cache using mechanisms such as multicast. They argue for a fixed hierarchy (i.e., a fixed parent-child relationship between proxies). In addition to consistency, they also consider pushing of content from origin servers to proxies. Mechanisms for scaling leases are studied in [68]. The approach assumes volume leases, where each lease represents *multiple objects* cached by a stand-alone proxy. They examine issues such as delaying invalidations until lease renewals and discuss prefetching and pushing lease renewals.

Another effort describes *cooperative consistency* along with a mechanism, called cooperative leases, to achieve it [52]. Cooperative consistency enables proxies to cooperate with one another to reduce the overheads of consistency maintenance. By supporting delta consistency semantics and by using a single lease for multiple proxies, the cooperative leases mechanism allows the notion of leases to be applied in a scalable manner to CDNs. Another advantage of the approach is that it employs application-level multicast to propagate server notifications of modifications to objects, which reduces server overheads. Experimental results show that cooperative leases can reduce the number of server messages by a factor of 3.2 and the server state by 20% when compared to original leases, albeit at an increased proxy-proxy communication overhead.

Finally, numerous studies have focused on specific aspects of cache consistency for content distribution. For instance, piggybacking of invalidations [40], the use of deltas for sending updates [48], an application-level multicast framework for Internet distribution [26] and the efficacy of sending updates versus invalidates [22].

References

[1] V. Almeida, A. Bestavros, M. Crovella, and A. de Oliveira. Characterizing reference locality in the WWW. In *Proceedings of PDIS'96: The IEEE Conference on Parallel and Distributed Information Systems*, Miami Beach, Florida, December 1996.

[2] The Apache Project. The Apache WWW server. http://httpd.apache.org.

[3] M. F. Arlitt and T. Jin. Workload characterization of the 1998 world cup web site. *IEEE Network*, 14(3):30–37, May/June 2000.

[4] M. F. Arlitt and C. L. Williamson. Internet Web servers: Workload characterization and performance implications. *IEEE/ACM Transactions on Networking*, 5(5):631–646, Oct 1997.

[5] G. Banga, J. Mogul, and P. Druschel. A scalable and explicit event delivery mechanism for UNIX. In *Proceedings of the USENIX 1999 Technical Conference*, Monterey, CA, June 1999.

[6] A. Barbir, B. Cain, F. Douglis, M. Green, M. Hofmann, R. Nair, D. Potter and O. Spatscheck. Known CDN request-routing mechanisms. IETF Internet-Draft, February 2002.

[7] P. Barford, A. Bestavros, A. Bradley, and M. E. Crovella. Changes in Web Client Access Patterns: Characteristics and Caching Implications. *World Wide Web Journal*, 1999.

[8] M. Beck and T. Moore. The Internet2 Distributed Storage Infrastructure Project: An Architecture for Internet Content Channels. In *Proceedings of the 3rd International Web Caching Workshop*, 1998.

[9] T. Berners-Lee, R. Fielding, and H. Frystyk. Hypertext transfer protocol – HTTP/1.0. IETF RFC 1945, May 1996.

[10] T. Brisco. DNS Support for Load Balancing. IETF RFC 1794, April 1995.

[11] P. Cao and C. Liu. Maintaining Strong Cache Consistency in the World-Wide Web. In *Proceedings of the Seventeenth International Conference on Distributed Computing Systems*, May 1997.

[12] V. Cardellini, M. Colajanni, and P. Yu. DNS Dispatching Algorithms with State Estimators for Scalable Web Server Clusters. *World Wide Web*, 2(2), July 1999.

[13] V. Cardellini, M. Colajanni, and P. Yu. Dynamic Load Balancing on Web-Server Systems. *IEEE Internet Computing*, pages 28–39, May/June 1999.

[14] V. Cate. Alex: A Global File System. In *Proceedings of the 1992 USENIX File System Workshop*, pages 1–12, May 1992.

[15] J. Challenger, A. Iyengar, K. Witting, C. Ferstat, and P. Reed. A Publishing System for Efficiently Creating Dynamic Web Content. In *Proceedings of IEEE INFOCOM 2000*, March 2000.

[16] M. Crovella and A. Bestavros. Self-similarity in World Wide Web traffic: Evidence and possible causes. *IEEE/ACM Transactions on Networking*, 5(6):835–846, Nov 1997.

[17] C. R. Cunha, A. Bestavros, and M. E. Crovella. Characteristics of www client-based traces. Technical Report CS 95-010, Boston University Computer Science Department, Boston, MA, June 1995.

[18] M. Day, B. Cain, G. Tomlinson, and P. Rzewski. A model for content internetworking (CDI). Internet Draft (draft-ietf-cdi-model-01.txt), February 2002.

[19] D. Dias, W. Kish, R. Mukherjee, and R. Tewari. A Scalable and Highly Available Web Server. In *Proceedings of the 1996 IEEE Computer Conference (COMPCON)*, February 1996.

[20] A. Downey. The structural cause of file size distributions. In *Proceedings of the Ninth International Symposium on Modeling, Analysis and Simulation of Computer and Telecommunication Systems (MASCOTS)*, Cincinnati, OH, Aug 2001.

[21] V. Duvvuri, P. Shenoy, and R. Tewari. Adaptive Leases: A Strong Consistency Mechanism for the World Wide Web. In *Proceedings of the IEEE Infocom'00, Tel Aviv, Israel*, March 2000.

[22] Z. Fei. A Novel Approach to Managing Consistency in Content Distribution Networks. In *Proceedings of the 6th Workshop on Web Caching and Content Distribution, Boston, MA*, June 2001.

[23] Z. Fei, S. Bhattacharjee, E. Zegura, and M. Ammar. A Novel Server Selection Technique for Improving the Response Time of a Replicated Service. In *Proceedings of IEEE INFOCOM'98*, 1998.

[24] R. Fielding, J. Gettys, J. Mogul, H. Frystyk, and T. Berners-Lee. Hypertext transfer protocol – HTTP/1.1. IETF RFC 2068, January 1997.

[25] R. Fielding, J. Gettys, J. Mogul, H. Frystyk, L. Masinter, P. Leach, and T. Berners-Lee. Hypertext transfer protocol – HTTP/1.1. IETF RFC 2616, June 1999.

[26] P. Francis. Yoid: Extending the Internet Multicast Architecture. Technical report, AT&T Center for Internet Research at ICSI (ACIRI), April 2000.

[27] C. Gray and D. Cheriton. Leases: An Efficient Fault-Tolerant Mechanism for Distributed File Cache Consistency. In *Proceedings of the Twelfth ACM Symposium on Operating Systems Principles*, pages 202–210, 1989.

[28] M. Gritter and D R. Cheriton. An Architecture for Content Routing Support in the Internet. In *Proceedings of the USENIX Symposium on Internet Technologies, San Francisco, CA*, March 2001.

[29] J. Gwertzman and M. Seltzer. World-Wide Web Cache Consistency. In *Proceedings of the 1996 USENIX Technical Conference*, January 1996.

[30] J. C. Hu, S. Mungee, and D. C. Schmidt. Techniques for developing and measuring high-performance Web servers over ATM networks. In *Proceedings of the Conference on Computer Communications (IEEE Infocom)*, San Francisco, CA, Mar 1998.

[31] J. C. Hu, I. Pyarali, and D. C. Schmidt. Measuring the impact of event dispatching and concurrency models on Web server performance over high-speed networks. In *Proceedings of the 2nd Global Internet Conference (held as part of GLOBECOM '97)*, Phoenix, AZ, Nov 1997.

[32] G. Hunt, G. Goldszmidt, R. King, and R. Mukherjee. Network Dispatcher: A Connection Router for Scalable Internet Services. In *Proceedings of the 7th International World Wide Web Conference*, April 1998.

[33] A. Iyengar and J. Challenger. Improving Web Server Performance by Caching Dynamic Data. In *Proceedings of the USENIX Symposium on Internet Technologies and Systems*, December 1997.

[34] A. Iyengar, J. Challenger, D. Dias, and P. Dantzig. High-Performance Web Site Design Techniques. *IEEE Internet Computing*, 4(2), March/April 2000.

[35] K. L. Johnson, J. F. Carr, M. S. Day, and M. F. Kaashoek. The measured performance of content distribution networks. In *International Web Caching and Content Delivery Workshop (WCW)*, Lisbon, Portugal, May 2000. http://www.terena.nl/conf/wcw/Proceedings/S4/S4-1.pdf.

[36] P. Joubert, R. King, R. Neves, M. Russinovich, and J. Tracey. High-performance memory-based web servers: Kernel and user-space performance. In *Proceedings of the USENIX Annual Technical Conference*, Boston, MA, June 2001.

[37] M. Koletsou and G. M. Voelker. The Medusa proxy: A tool for exploring user-perceived web performance. In *Proceedings of International Web Caching and Content Delivery Workshop (WCW)*, Boston, MA, June 2001. Elsevier.

[38] B. Krishnamurthy and J. Rexford. *Web Protocols and Practice*. Addison Wesley, 2001.

[39] B. Krishnamurthy and C. Wills. Proxy Cache Coherency and Replacement—Towards a More Complete Picture. In *Proceedings of the 19th International Conference on Distributed Computing Systems (ICDCS)*, June 1999.

[40] B. Krishnamurthy and C. Wills. Study of Piggyback Cache Validation for Proxy Caches in the WWW. In *Proceedings of the 1997 USENIX Symposium on Internet Technology and Systems, Monterey, CA*, pages 1–12, December 1997.

[41] B. Krishnamurthy, C. Wills, and Y. Zhang. On the use and performance of content distribution networks. In *Proceedings of ACM SIGCOMM Internet Measurement Workshop*, November 2001.

[42] T. T. Kwan, R. E. McGrath, and D. A. Reed. NCSA's World Wide Web Server: Design and Performance. *IEEE Computer*, 28(11):68–74, November 1995.

[43] D. Li, P. Cao, and M. Dahlin. WCIP: Web Cache Invalidation Protocol. IETF Internet Draft, November 2000.

[44] B. Mah. An empirical model of HTTP network traffic. In *Proceedings of the Conference on Computer Communications (IEEE Infocom)*, Kobe, Japan, Apr 1997.

[45] Z. Morley Mao, C. D. Cranor, F. Douglis, M. Rabinovich, O. Spatscheck, and J. Wang. A precise and efficient evaluation of the proximity between web clients and their local DNS servers. In *Proceedings of USENIX Annual Technical Conference*, June 2002.

[46] J. C. Mogul. Clarifying the fundamentals of HTTP. In *Proceedings of WWW 2002 Conference*, Honolulu, HA, May 2002.

[47] J. C. Mogul. Network behavior of a busy Web server and its clients. Technical Report 95/5, Digital Equipment Corporation Western Research Lab, Palo Alto, CA, October 1995.

[48] J C. Mogul, F. Douglis, A. Feldmann, and B. Krishnamurthy. Potential Benefits of Delta Encoding and Data Compression for HTTP. In *Proceedings of ACM SIGCOMM Conference*, 1997.

[49] D. Mosedale, W. Foss, and R. McCool. Lessons Learned Administering Netscape's Internet Site. *IEEE Internet Computing*, 1(2):28–35, March/April 1997.

[50] E. M. Nahum, T. Barzilai, and D. Kandlur. Performance issues in WWW servers. *IEEE/ACM Transactions on Networking*, 10(2):2–11, Feb 2002.

[51] E. M. Nahum, M. Rosu, S. Seshan, and J. Almeida. The effects of wide-area conditions on WWW server performance. In *Proceedings of the ACM Sigmetrics Conference on Measurement and Modeling of Computer Systems*, Cambridge, MA, June 2001.

[52] A. Ninan, P. Kulkarni, P. Shenoy, K. Ramamritham, and R. Tewari. Cooperative Leases: Scalable Consistency Maintenance in Content Distribution Networks. In *Proceedings of the World Wide Web conference (WWW2002)*, May 2002.

[53] Open Market. FastCGI. http://www.fastcgi.com/.

[54] V. N. Padmanabhan and L. Qui. The content and access dynamics of a busy web site: findings and implications. In *SIGCOMM*, pages 111–123, 2000.

[55] V. Pai, M. Aron, G. Banga, M. Svendsen, P. Druschel, W. Zwaenepoel, and E. M. Nahum. Locality-Aware Request Distribution in Cluster-based Network Services. In *Proceedings of ASPLOS-VIII*, October 1998.

[56] V. Pai, P. Druschel, and W. Zwaenepoel. Flash: An efficient and portable Web server. In *USENIX Annual Technical Conference*, Monterey, CA, June 1999.

[57] V. S. Pai, P. Druschel, and W. Zwaenepoel. I/O Lite: A copy-free UNIX I/O system. In *3rd USENIX Symposium on Operating Systems Design and Implementation*, New Orleans, LA, February 1999.

[58] M. Rabinovich and O. Spatscheck. *Web Caching and Replication*. Addison-Wesley, 2002.

[59] A. Shaikh, R. Tewari, and M. Agrawal. On the Effectiveness of DNS-based Server Selection. In *Proceedings of IEEE INFOCOM 2001*, 2001.

[60] J. Song, A. Iyengar, E. Levy, and D. Dias. Architecture of a Web Server Accelerator. *Computer Networks*, 38(1), 2002.

[61] The Standard Performance Evaluation Corporation. SpecWeb99. http://www.spec.org/osg/web99, 1999.

[62] *Squid Internet Object Cache Users Guide*. Available on-line at http://squid.nlanr.net, 1997.

[63] R. Srinivasan, C. Liang, and K. Ramamritham. Maintaining Temporal Coherency of Virtual Warehouses. In *Proceedings of the 19th IEEE Real-Time Systems Symposium (RTSS98), Madrid, Spain*, December 1998.

[64] Sun Microsystems Inc. The Java Web server. http://wwws.sun.com/software/jwebserver/index.html.

[65] R. Tewari, T. Niranjan, and S. Ramamurthy. WCDP: Web Content Distribution Protocol. IETF Internet Draft, March 2002.

[66] Red Hat Inc. The Tux WWW server. http://people.redhat.com/ mingo/TUX-patches/.

[67] D. C. Verma. *Content Distribution Networks: An Engineering Approach*. John Wiley & Sons, 2002.

[68] J. Yin, L. Alvisi, M. Dahlin, and A. Iyengar. Engineering Server-driven Consistency for Large-scale Dynamic Web Services. In *Proceedings of the 10th World Wide Web Conference, Hong Kong*, May 2001.

[69] J. Yin, L. Alvisi, M. Dahlin, and C. Lin. Volume Leases for Consistency in Large-Scale Systems. *IEEE Transactions on Knowledge and Data Engineering*, January 1999.

[70] J. Yin, L. Alvisi, M. Dahlin, and C. Lin. Hierarchical Cache Consistency in a WAN. In *Proceedings of the Usenix Symposium on Internet Technologies (USITS'99), Boulder, CO*, October 1999.

[71] H. Yu, L. Breslau, and S. Shenker. A Scalable Web Cache Consistency Architecture. In *Proceedings of the ACM SIGCOMM'99, Boston, MA*, September 1999.

[72] Zeus Inc. The Zeus WWW server. http://www.zeus.co.uk.

Network Management: State of the Art

Raouf Boutaba and Jin Xiao
Department of Computer Science
University of Waterloo, CANADA
Email: *{rboutaba, j2xiao}@bbcr.uwaterloo.ca*

Abstract: This paper examines the state-of-the-art enabling technologies for network management, including policy-based network management, distributed object computing, Web-based network management, Java-based network management, code mobility, intelligent agents, active networks, and economic theories. For each of them, we discuss the underlying concept, analyze the benefits and drawbacks, and discuss the applicability to network management. In doing so, we illustrate the common trend in network management design: moving towards distributed intelligence.

1 INTRODUCTION

Nearly a decade ago, the classic agent-manager centralized paradigm was the pervasive network management architecture, exemplified in the OSI reference model, the Simple Network Management Protocol (SNMP) management framework, and the Telecommunications Management Network (TMN) management framework [15]. With the increasing size, management complexity, and service requirement of today's networks, such management paradigm is no longer adequate, and should be replaced with distributed management paradigms. This trend is clearly discussed in [29]. With the myriads of enabling technologies surfaced in the last few years, all of which offering various degrees of network management distribution and benefits, it is unclear what, when, and where are these technologies most applicable? And what would the future of network management be? By examining these state-of-the-art enabling technologies, this paper attempts to shed some light on their benefits, drawbacks, and postulate on their future prospects in network management. Despite their diversity, the paper will illustrate a recurring trend in their design concept: pushing towards distributed intelligence. In a nutshell, management agents are no longer treated as "dumb terminals", but as sophisticated computing devices, and are exploited as such. Distributed intelligence denotes the management capability and autonomy a management agent exhibits.

This paper is comprised of three parts. First, we will briefly outline the network management concepts, its objectives, and the unique challenges future network management brings. Then we will examine the key enabling technologies for network management. Lastly, we will compare these technologies in terms of distributed intelligence and network resource consumption.

2 NETWORK MANAGEMENT: OBJECTIVES AND CHALLENGES

Hegering [13] defines network management as all measures ensuring the effective and efficient operations of a system within its resources in accordance with corporate goals. To achieve this, network management is tasked with controlling network resources, coordinating network services, monitoring network states, and reporting network status and anomalies. In our view, the objectives of network management are:

- **Managing network resources and services:** including the control, monitor, update, and report of network states, device configurations, and network services.
- **Simplify network management complexity:** it is the task of network management systems to extrapolate network management information into human manageable form. Conversely, network management systems should also have the ability to interpret high-level management objectives.
- **Reliable services:** to provide network with high quality of service, minimize network downtime. Network management systems should detect and fix network faults and errors. And network management must safeguard against all security threats.
- **Cost conscious:** Network management should keep track of network resources and network users. All network resource and service usage should be tracked and reported.

OSI has a well-defined network management reference model [14] pertinent to the designs of current network management architectures. The OSI model breaks network management functions into the following five functional areas:

- **Fault Management:** the detection, recovery, and documentation of network anomalies and failures.
- **Configuration Management:** record and maintain network configuration, update configuration parameters to ensure normal network operations.

- **Accounting Management:** user management and administration, billing on usage of network resources and services.
- **Performance Management:** provide reliable and high quality network performance. This includes quality of service provisioning and regulating crucial performance parameters such as network throughput, resource utilization, delay, congestion level, and packet loss.
- **Security Management:** provide protection against all security threats to network resources, its services, and data. In addition, ensure user privacy and control user access rights.

In the recent years, network infrastructure is shifting towards service-centric networks. Besides the above network management objectives and OSI functional areas, network management must also fulfill additional management requirements, similar to today's business service models: fast time to market, service differentiation, service customizability, more features, and flexibility.

We envision the future of network infrastructure will drastically change the way network management is done and presents new challenges to network management. First of all, as the size of networks continue to grow at current rate, more and more network devices need to be managed efficiently, demanding better scalability on network management designs. As a result of such size increase, human directives can only be given at a very high level of abstraction and generalization. The underlying network management system must take care of the interpretation of these high-level directives to realizable network configurations and oversee their enforcement. Secondly, as network infrastructures from various sectors converge, heterogeneous network technologies must co-exist and inter-work. Network management systems must provide such seamless integration via common service interfaces, and hide underlying technological heterogeneity from network users. Thirdly, the competitive nature of current network services demands economical operation of networks. Network management must also be more self-regulating and self-governing, in order to be economically beneficial. At the same time, network management solutions must be kept simple and elegant, as the development of Internet has demonstrated: only simple and elegant solutions would prevail in large-scale heterogeneous networks. Lastly, as network devices become more and more powerful, there is increasing pressure to utilize their processing capabilities. This leads to increasing need for distributed network management at device level.

3. EARLY WORKS TOWARDS DISTRIBUTED NETWORK MANAGEMENT

In the traditional manger-agent network management architecture, such as SNMP, the agent is kept as simple as possible, only tasked with device status report and update, while the burden of management and data processing resides with the manager. Researchers realized the inadequacy of such design around early 90's, as the rapid increase in size of managed network, compounded by increasing demand on network performance and reliability, prompted a complete re-thinking of network management paradigm.

SNMPv2 is the first major installment towards distributed network management. The initial set of Request For Comments (RFCs) (1441-1452) was published in 1992. SNMPv2 introduced the concept of intermediary manager. An intermediary manager can be viewed as a "middle manager". The manager communicates directly with the intermediary managers and exchange command information, while the intermediary managers handle data exchange with agents. In this fashion, the intermediary managers shift some of the data processing from the manager side and is capable of performing simple tasks, such as periodic status pulling from agents, without manager's direct intervention.

In 1995, Internet Engineering Task Force (IETF) took a further step towards management distribution with the proposal on Remote MONitoring (RMON) [38]. RMON used the concept of monitors or probes, which are network traffic monitoring devices. Probe implementation can be done as device embedded applications or as separate devices. The task of a probe is to monitor the network traffic at its local region and report anomalies, in the form of alarms, to its manager. By defining alarm types and alarm thresholds, the manager is able to offload some data gathering and decision-making (mainly event filtering) to the probes. Furthermore, the probes can also perform some data pre-processing before forwarding them to the manager.

In general, the earlier works towards distributed network management can be considered as weak distribution. The management tasks still reside heavily on the manager side, and some rudimentary management duties are delegated to intermediary entities, in the form of event filtering, notification, and data pre-processing.

4. ENABLING TECHNOLOGIES

We have identified a set of enabling technologies that are commonly recognized to be potential candidates for distributed network management. We will discuss each of them in turn, examine their potential benefits to network management, discuss their drawbacks, and postulate on their prospects. These enabling technologies will be presented in order, with respect to the degree of management capability it bestows on management agents. We believe that distributing intelligence to management agents is an inevitable trend in network management and one that is critical to the success of future network management designs. We will first examine policy based network management. It will be followed by distributed computing, Web-based systems, and Java, which all uses static remote objects to facilitate task offloading from agent to managers. From there, we present the concept of code mobility, in which agents are more management capable, as agents are made mobile and exhibit the ability of independent management processing. A step further in that direction is intelligent agents, where processing units cooperate with each other on peer-to-peer basis, assuming the role of managers and agents interchangeably. Lastly, we will examine the application of active network and economic theories to network management. The former pushes management tasks completely to network devices, and the later forgoes the need for network management infrastructure.

4.1 Policy-based Network Management

Policy-based network management started in early 1990s [30][23]. Although the idea of policies appears even earlier, they were used primarily as representation of information in a specific area of network management: security management [11]. The idea of policy comes quite naturally to any large management structures. In reality, all medium to large size companies today have policies and regulations that their employees must follow. These policies are typically derived based on company's objectives and goals. In policy-based network management, policies are defined as rules that govern the states and behaviors of the network system. The management system is tasked with: the transformation of human-friendly management goals to syntactical and verifiable rules governing the function and status of the network, the translation of such rules to mechanical and device-dependent configurations, and the distribution and enforcement of these configurations by management entities. The reference model of policy-based network management is largely a manager-agent model, consists of Policy Decision Points (PDPs) and Policy Enforcement Points (PEPs) [18][19]. The first two tasks are handled by the PDPs, while the last task is handled by the PEPs.

IETF's Resource Allocation Protocol (RAP) plays a key role in policy-based network management with its Common Open Policy Services (COPS) [20] and its extension COPS-PR [9]. Some recent works are done on the translation of business directives to network level policies [8] and on policy conflicts resolution [27]. More significantly, the meta-policies concept was proposed in [3]. Its introduction pushes most mundane policy decision tasks from the PDPs to the PEPs. This represents a novel attempt at empowering agents with more management capabilities, moving policy-based network management towards a more distributed intelligence design.

The most significant benefit of policy-based network management is that it promotes the automation of establishing management level objectives over wide-range of network devices. Network administrator would interact with the network by providing high-level abstract policies. Such policies are device independent and human-friendly. The automated translation process will hide the complexity of constructing low-level device-dependent configurations derived from the high-level policies, and therefore facilitate the bridging of business objectives to network configurations. Comparing to human-directed policy translation, such automation would provide more consistent and integrated representation of business objectives. As the state of a network changes, policies would be automatically updated to ensure operational consistency without any human interventions. As today's network increases rapidly in size, such automation is an essential requirement. In contrast to other management technologies, such as Java-based management and mobile agent, policy-based network management allows much more rapid modification of the management requirements after deployment. Policy-based network management can adapt rapidly to changing management requirements via run-time reconfigurations, rather than re-engineer new object modules for deployment. The introduction of new polices does not invalidate the correct operation of a network, provided the newly introduced polices does not conflict with existing policies. In comparison, a newly engineered object module must be tested thoroughly in order to obtain the same assurance.

For large networks with frequent changes in operational directives, policy-based network management offers an attractive solution, as it can dynamically translate and update high-level business objectives into realizable network configurations. However, one of the key issues in a policy-based network management lies in its functional rigidity. After the development and deployment of a policy-based network management system, the service primitives are defined. By altering management policies and modifying constraints, we have a certain degree of flexibility in cooping with changing management directives. However, we cannot modify or add new management services to the system, unlike mobile code or software agents.

4.2 Distributed Object Computing

Distributed Object Computing (DOC) uses Object-Oriented (OO) methodology to construct distributed applications. Its adaptation to network management is aimed at providing support for distributed network management architecture, integration with existing heterogeneous network management solutions, and provide development tools for distributed network management components. Distributed object computing provides distribution of services and applications in a seamless and location transparent way, by separating object distribution complexity from network management functionality concerns. Another advantage of this separation of concerns is the ability to provide multiple management communication protocols accessed via a generalized Abstract Programming Interface (API), fostering interoperability of heterogeneous network management protocols, such as SNMP for IP networks and Common Management Information Protocol (CMIP) for telecommunication networks. In addition, DOC provides distributed development platform for rapid implementation of robust, unified, and reusable services and applications. Contemporary DOC in network management is oriented around the Object Request Broker (ORB) concept. ORB facilitates communication between local and remote objects in an effortless way that free the application from low-level infrastructure and communication concerns. The two major adaptation of DOC to network management are: Common Object Request Broker Architecture (CORBA) [32] and Distributed COM (DCOM) [34].

The major application of DOC to network management is mostly in two areas. Firstly, DOC is used to design distributed network management systems, evident in standardization works done by Telecommunication Information Network Architecture Consortium (TINA-C) [33], Joint Inter Domain Management (JIDM) [16], and research projects, such as MESIS [2]. All of these proposed frameworks provide transparent remote services invocation using DOC support. In this fashion, management processing and services need no longer be located at centralized locations in the network, but rather distributed across remote locations. This feature allows management tasks to be delegated, by region or by functional areas, to intermediate entities, making managers no longer the center of all management decision making. Secondly, DOC is used to augment existing network management infrastructures with distributed capability.

Distributed object computing in general, CORBA in particular, is a well-received technology for developing integrated network management architectures with object distribution. The success of CORBA as an enabling network management technology can be attributed to the fact that CORBA has well-established supporting environment for efficient run-time object distribution and a set of support services. In this fashion, DOC is useful as integration tools for heterogeneous network management domains, and extending deployed network management architectures. However, DOC still uses static object distribution. It does not have the flexibility code mobility offers. Furthermore, DOC requires dedicated and heavy run-time support, which may not always be feasible on every device in the network. This later issue restricts its area of deployment.

4.3 Web-based Network Management

Judging by the tremendous success of World Wide Web on the Internet, it is expected that web technology would influence network management to some degree. Today, myriad of web-based network management solutions are proposed and been built, backed up by large corporations, such Sun, Cisco, Microsoft, etc. With respect to network management, the critical problems Web-based network management tries to address are: platform heterogeneity, lack of management console accessibility, and high cost of management platform deployment and maintenance. Traditional network management solutions are highly platform-dependent. Network administrators must operate on proprietary management consoles to perform daily operations, and the user interfaces for each management platform may vary significantly. Web technology effectively addresses this problem by providing ubiquitous management consoles in the form of standard web browsers. Proprietary network management platforms are expensive and difficult to maintain. Web technology solves this issue by promoting HyperText Markup Language (HTML) and Java applet in information presentation, providing a seamless Graphic User Interface (GUI) accessible everywhere. Lastly, an interesting observation in the IP sector is that network management data is always treated as "second class citizens" compare to user data. While it's true that the transport of management data should never get in the way of transporting user data, the importance of management data is on the rise, especially with the increasing demand on real-time Quality of Service (QoS) services. Using a connection-oriented transport protocol, such as Transport Control Protocol (TCP) for HyperText Transport Protocol (HTTP), implicitly elevates management data to the same level as user data, as viewed by network routers. Web technology serves as a good short-term solution to "patch" the existing issues in network management, as new management paradigms mature, which would take quite sometime to develop and standardize.

We define web infusion as the degree to which web technology is incorporated into a network management platform, ranging from platform extension, to component modification, to full web-based management platforms. In our view, there are three degrees of web infusion existing today: web gateways, web-embedded servers, and web-based management platforms. The web gateways are independent components situated in between web browser type management consoles and management agents, which are implemented as various platform-dependent entities, such as SNMP agents. The web gateway is responsible for the translation of HTTP request to SNMP/CMIP request, and the formulation of web documents based on data gathered from managed devices. The web gateway is extremely easy to deploy, since it does not require any modification on existing management architectures. However, its development can be complex, since it is, by nature, a multi-protocol architectural gateway. In large networks, the presence of web gateways may become performance bottlenecks, as all requests to managed devices have to go through these gateways. The web-embedded servers apply web technology to all managed devices, such as presented in [21] [28]. Each managed device is a miniature web server, capable of accepting HTTP request, processing device data, constructing HTML/XML presentation of device data, and transmitting constructed documents. Because of the self-contained nature of web-embedded servers, there is no requirement for additional management support. A network administrator can simply interact with a web-embedded device via standard web browser. However, web-embedded servers are not deployable on devices with limited resources and processing power, as it leaves relatively large network footprints. In addition, there are no efficient and economical methods of transforming existing network devices into web-embedded servers. In contrast, Web-based management platforms use web technology as the core technology in the design of new network management platforms, with its own management protocol, data model, and architecture. Web-Base Enterprise Management (WBEM) [37] is a well-known example of web-based management platform. The first two types of web infusion are by far the most adopted solutions in the network management domain today. In both cases, preliminary processing of device data, formulation of status report, and GUI presentation are handled by separate entities other than network managers.

Recently, there has been much debate over the right technology for integrated network management. Since both web technology and CORBA are widely used for this purpose, the question posed comes as no surprise. At first sight, web does seem to be a better choice, as many web advocates believe. Web technology removes the need for proprietary management consoles; it provides uniform management information access via web browsers; data modeling in HTML form is easier than defining Interface Definition Languages (IDLs); with the exception of embedded web servers, web-based management does not need dedicated runtime environment and leaves very small network device footprint; web technology has matured security measures that can be exploited; HTTP based data transport is inherently reliable. However, as we examine the inner works of web technology more closely, the strength of CORBA becomes apparent. Web-based management usually involves much runtime interpretation, in terms of HTML/XML documents, CGI/SSI scripts, and Java applets. These runtime interpretations are a cause of performance concerns, especially for real-time control. HTML/XML are constructed for human readability, hence the formats of these documents tend to be overly wordy for representing key-value pairs, which are the most common type of information in network devices. CORBA's IDL would be more compact for these types of data representation. And this compactness translates directly to network bandwidth savings. By using web technology, the developers are limited to using TCP transport for management data, which may or may not be the best choice. CORBA does not place this restriction on the developers. Lastly, CORBA inherently supports distributed management paradigm, by providing support for distributed object development and object distribution transparency. Web technology does not make implementing distributed paradigm in network management any easier. The burden of implementing distribution is largely left to higher-level management architecture. Overall, the choice of technology should be determined based on particular circumstances. In general, web-based technology is better used for providing web access to managed devices, especially if the user of the management application does not have much domain-specific knowledge, e.g. Customer-directed network resource configuration. CORBA is best used for fully distributed network management platforms that values operational efficiency over accessibility. Of course, the two technologies can also be combined in the same management platforms, whereby the web technology could offer access to CORBA-based management applications and services.

4.4 Java-based Network Management

Java, being a portable and object-oriented programming language, is the instrumentation for a wide variety of network management paradigms, ranging from distributed computing, to web-based management, to intelligent agents. Because of this wide applicability, many Java-based development environments have been proposed and designed, supporting network management applications. What makes Java a good technology for network management in general? Firstly, deploying Java-based software solutions are relatively cheap compare to other management software solutions, such as CORBA-based applications. Java virtual machine (JVM) is the only runtime support needed by a Java-based software, and it is also easily deployable and requires very little maintenance. Secondly, as more and more JVM-enabled network devices become available, so does the availability of java support. Furthermore, Java can interoperate with web browsers, which are good candidates for cheap and accessible management consoles. Thirdly, dynamic code downloading allows dynamic distribution of java objects. This not only opens the opportunity for runtime service extensions, but also opens the opportunity for management delegations. Fourth, Java is platform-independent, portable on any existing management platforms that support JVM. Lastly, Java software is easy to develop, as there exists many development supporting environment and tools. Also, Java is a good programming language for realizing new network management concepts, such as code mobility.

Perhaps the biggest and most mentioned issue with Java is its performance. Java is inherently not an efficient programming language. Besides the obvious performance loss resulting from Java's interpreted nature, Java class loading can be quite slow, especially if dynamic class downloading is required. Java object serialization and remote method invocation are commonly exploited for network management. Both of them have performance problems. Object serialization is quite space consuming, which may not be a big problem on large stations, but would be an issue for small devices. Java Remote Method Invocation (RMI) is not network resource conscious in its operation and tends to waste fair amount of network resources on each method invocation.

4.5 Code Mobility for Network Management

In 1991, Yemini et al. first introduced the concept of Management by Delegation (MbD), and they further refined this concept in 1995 [12]. In their works, Yemini et al. suggested to push management tasks to the agent side. This can be achieved by dynamically transporting programs from managers to agents and perform the delegated management tasks locally. Three immediate advantages of the MbD approach are apparent. Firstly, manager is no longer a centralized processing entity in the network. Much of its processing can be offloaded to agents via delegated programs. Secondly, considerable amount of network resources are saved. For instance, data gathering can be performed locally. Lastly, It is possible to augment the functionality of agents by providing them with delegated programs at runtime. In this fashion, some decision making and network monitoring duties can be performed locally, allowing faster response to management requests and better fault tolerance (in case of manager crash).

Code mobility can be considered as the capability of an application to distribute and relocate its components at run-time. Obviously, there must exist some form of language and run-time support for applications utilizing code mobility. There is much confusion in the literature concerning the terminologies used for code mobility, and the introduction of intelligent agents further blurs the concept. We do not consider intelligent agent as part of the code mobility concept in this paper. Hence, intelligent agents are considered more complex and self-governed than code mobility, and will be discussed in a separate section. In terms of code mobility, there exist two types: weak mobility and strong mobility. In weak mobility, entire programs or code fragments are transported between distributed components, without retaining execution states and data after transportation. We call applications exhibiting weak mobility as mobile code. Recent works such as [25] explores its use for network management. In strong mobility, the entire program, along with its execution states and data, are transported between remote components. The program will suspend its execution before departure and resumes execution after arrival. We call applications exhibiting strong mobility as mobile agent. Most research works, such as [35][22][26] are focused on this concept. The terms mobile code and mobile agent are often used interchangeably, and sometimes mean different things across literatures.

With code mobility, management tasks no longer have to be performed by the managers. They simply generate management objectives and outline task procedures, the execution of tasks are delegated to the agents. Baldi and Picco [1] defined three code-mobility paradigms based on interaction between services and resources: Code On Demand (COD), Remote Evaluation (REV), and Mobile Agent (MA). In the case of code on demand, the manager has gathered the resources but lacks the code needed for processing. The code is dynamically downloaded from a code server for execution. In the case of remote evaluation, the manager holds the code and the agent holds the resources. The manager dynamically uploads code to the agent side. The uploaded code executes on the resources, and returns back the result to the manager. In the case of mobile agent, the manager holds the services in the form of processing components and the agent holds the resources. The manager relocates the entire processing component, which includes code, execution state, and possibly data, to the agent. If the required data is distributed across a number of different agents, the mobile agent has the ability to relocate from agent to agent, performing data processing and keeping track of generated intermediary data. The MA paradigm is characteristic of strong mobility, while COD and REV paradigms are characteristic of weak mobility.

As mobile code is transported across network, it must be loaded at the destination for execution. The time it takes to suspend execution of a component, pack its code and data, transport across network, restore the component, and execute, could be quite long. Hence, code mobility is not a good candidate for networks with simple but frequent service requests. Furthermore, to prevent mobile agents from adversely affecting network resources, security measure are often in place which either restrict the operations a mobile agent can perform on local resources, or provide some type of access gateway. Neither solution is satisfactory as access restrictions constrain the operational capacity of mobile agents; while access gateways add unnecessary processing overhead.

4.6 Intelligent Agents

Intelligent agents exhibit the following characteristics: autonomy, social ability, reactivity, pro-activeness, mobility, learning, and beliefs. An intelligent agent is an independent entity capable of performing complex actions and resolving management problems on its own. Unlike code mobility, an intelligent agent does not need to be given task instructions to function, rather just the high-level objectives. The use of intelligent agents completely negates the need for dedicated manager entities, as intelligent agents can perform the network management tasks in a distributed and coordinated fashion, via inter-agent communications. Many researchers believe intelligent agents are the future of network management, since there are quite some significant advantages in using intelligent agents for network management. Firstly, intelligent agents would provide a fully scalable solution to most areas of network management. Hierarchies of intelligent agents could each assume a small task in its local environment and coordinate their efforts globally to achieve some common goal, such as keeping overall network utilization at close to maximum. Secondly, data processing and decision-making are completely distributed, which alleviates management bottlenecks as seen in centralized network management solutions. In addition, the resulting network management system is more robust and fault tolerant, as the malfunction of small number of agents have no significant impact on the overall management function. Thirdly, the entire network management system is autonomous, network administrators would only need to provide service-level directives to the system. Lastly, the intelligent agents are self-configuring, self-managing, and self-motivating. It is ultimately possible to construct a network management system that's completely self-governed and self-maintained. Such a system would largely ease the burden of network management routines that a network administrator has to currently struggle with.

Wooldridge and Jennings [39] defined three architectural types for intelligent agents: deliberative agents, reactive agents, and hybrid agents. Deliberative agents are based on a physical-symbol system. Such a system describes a physically realizable set of symbols that can be combined to form complex structures. A deliberative agent is able to run processes operating on these symbols to generate overall intelligent actions. Recent works such as [24] make use of deliberative agent. Reactive agents are very much the opposite of deliberative agents. They do not require complex representation of knowledge, nor do they require perfect representation of information. Reactive agents generate behaviors solely based on environmental observations, since they do not include any kind of symbolic world models. In practice, reactive agents are more responsive than deliberative agents due to the lack of any complex symbolic reasoning mechanism. Reactive agents could be successfully applied to traffic monitoring, fault diagnosis, congestion control, and admission control, because these management functions do not have or require perfect representation of a world model. Furthermore, they require rapid responses and actions, which the reactive agents are capable of. Hybrid agents are compositions of both deliberative agents and reactive agents. A hybrid agent would contain a symbolic world model, developing plans, and making decisions in the way a deliberative agent functions. However, it is also capable of reacting to events occurring in the environment without engaging in complex reasoning. The reactive component of a hybrid model overwrites its deliberative component in order to achieve quick response. The hybrid agent seems to be a suitable candidate for fault diagnosis [10]. However, hybrid agents are substantial in size, much larger than either deliberative agents or reactive agents. This may pose a problem when high levels of mobility are expected in a network management system.

The application of intelligent agents to network management is still at its infancy, and much difficult issues still remain unsolved. As applications utilizing intelligent agents arise in network management, the problem of managing these intelligent agents also becomes increasingly important. These self-governing agents cannot simply be allowed to roam around the network freely and access vital resources. Currently, it is still very difficult to design and develop intelligent agent platforms. This is mostly because very little real-life practices with intelligent agents exist today. We have yet to determine what constitutes a good intelligent agent platform, in practical terms. As more intelligence and capabilities are empowered to the intelligent agents, their size becomes an increasing concern for network transport. Furthermore, agent-to-agent communications typically uses Knowledge Query Manipulation Language (KQML). KQML wastes substantial amount of network resources, as its messages are very bulky. Lastly, protection against malicious intelligent agents is hardly addressed in the current literature. Who takes care of agent authentication? Can agents protect themselves against security attacks? Can agents keep their knowledge secret? How much access rights should agents have over network resources? None of these questions are addressed effectively, and until they do, large deployment of intelligent agents for network management is very unlikely.

4.7 Active Networks

According to Tennenhouse et al. [36], an active network is a new approach to network architecture in which the network nodes, such as routers and switches, perform customized computation on messages flowing through them. In active networks, routers and switches run customized services that are uploaded dynamically from remote code servers or from active packets. The characteristic of activeness is three folds. In device view, a device's services and operatives can be dynamically updated and extended actively at run-time. In network provider view, the entire network resources can be provisioned and customized actively on per customer basis. In network user view, the allocated resources can be configured actively based on user application needs.

Active networks, combined with code mobility, present an effective enabling technology for distributing management tasks to device level. Not only does management tasks can be offloaded to individual network devices, but also the supplier of management task need no longer be manager entities. Such a solution provides full customizability, device-wise, service provider-wise, and user-wise; it provides the means for distributed process across all network devices; it is interoperable across platforms via device-independent active code; it fosters user innovation and user-based service customization; it accelerates new service and network technology deployment, bypassing standardization process and vendor consensus; it allows for fine grained resource allocation based on individual service characteristics. In the literature, there are two general approaches for realizing active networks: programmable switch approach and capsule approach. Programmable switch approach uses out-of-band channel for code distribution. The transportation of active code is completely separated from regular data traffic. This approach is easier to manage and secure, as the active code is distributed via private and secure channels. It is suited for network administrators configuring network components. On the other hand, the capsule approach packages active code into regular data packets. The active code is sent to active node via regular data channel. This approach allows open customization of user-specified services, however, is more prone to security threats. [4] analyzed the benefits of active networks to enterprise network management.

Quite some recent works are done on exploring active networks for network management, such as the Virtual Active Network (VAN) proposal [7] and the agent-based active network architecture [17]. However, security remains a major roadblock for practical application of active network. Not only the integrity of network resources and user data has to be kept, but also the content of user data must remain confidential. The later implies a strong trust on the active nodes a packet must visit en-route to destination, as it is necessary for user data to be examined and processed in some form. As noted by Murphy et al. [31], there are many objects of security concern in active networks, including: end users, active nodes, Execution Environments (EEs), and active codes. The trust models for these objects are also quite complex.

Besides security, resource provisioning and fault tolerance are the other two major issues that need to be addressed in active networks. Firstly, as resources are used for customized processing of data packets in the network. Some means of governing the priority of resource access and the limit of resource consumption has to be established. This issue creates new requirements for network management that must be addressed. Another related issue is network bandwidth consumption. After all, user-specific services must be transported across the network and uploaded. If capsule approach is used, the transportation of these services comes in direct contention with the transportation of user data. Simply charging user for service deployment may not be desirable since it discourages the user from customizing the active nodes in the network. Secondly, fault tolerance of the network will suffer if user-specific services aren't controlled properly. As user gains the ability to manage network resources and perform customized processing, more and more user services/applications are injected into the network. The quality of these services/applications cannot be as well ascertained as the manufacturer-supplied services. The obvious solution is to providing each user service with a separate and isolated execution environment. However, such a solution is very costly in terms of resource consumption and network performance.

4.8 Economic Theory

Network management using economic theory proposes to model the network services as an open market model. The resulting network is self-regulating and self-adjusting, without the presence of any formal network management infrastructure. Network administrators would indirectly control the network dynamics by inducing incentives and define aggregate economic policies. Such an approach may seem to be very bold, but it draws its theory from the well-established economic sciences. The premises for applying economic theories are: the existence of open and heterogeneous networks; multi-vendor orientation; and competitive services. Very few works have been done on this subject matter, and most of them are focused on using economic theory as agent coordination model [5][6]. As discussed previously, the management of intelligent agents is still neglected in current literatures. Using economic theory for managing multi-agent systems could be a viable alternative, due to its simplicity and self-sustaining nature.

However, the application of economic theories to network management is only at early experimental stage. Many critical issues brought out with these experiments cast doubts on the applicability of economic theory to network management. Using market model for managing networks is a novel idea. However, some important design issues must be carefully considered. Firstly, the driving force for a market model is the authenticity of its currency. Hence currency values and its transaction processes used in market model must be secure. Furthermore, such secure transactions must be performed very efficiently, as it would be a very frequent operation. Secondly, economic policy for the market model must be designed in such a way that it encourages fair competition, and strongly relates resource contention and its associated price. Lastly, the market model would be operating on a wide scale, requiring standardization of its elements and operations. Such standardization may be a very slow process and would require full consensus from all participating vendors.

5 CONCLUSION

All of the enabling technologies discussed in this paper attempts to provide distributed intelligence to management agents. Policy-based network management allows managers to partially delegate management tasks to agents in the form of concrete policies. Web-network management offloads the processing, presentation, and display of device information to web gateways or embedded web servers. Distributed object computing, such as CORBA, and Java-based network management provides the means for management task distribution in the network, via deploying static distributed objects. Code mobility and active networks delegate management tasks to management agents through dynamic mobile code downloading. Intelligent agents push distributed intelligence even further by defining autonomous agents that are capable of making complex management decisions. The role of such intelligent agents is no longer confined to either the manager or the agent, as the intelligent agents can adopt these roles dynamically, based on their assigned tasks or their own motivations. Lastly, economic theories completely negate the need for a network management infrastructure, by modeling the network as a self-regulating open market.

To fully leverage the benefits of the presented enabling technologies, network management designers must balance all the benefits and drawbacks, as discussed in this paper. We believe that distributed intelligence is one of the most important trends in the management of current and future large-scale complex networks. Despite the diversity of these enabling technologies, their use in network management research aims at distributing intelligence in the network.

6 REFERENCES

[1] Baldi M., Picco G., Evaluating the Tradeoffs of Mobile Code Design Paradigms in Network Management Applications, 1998
[2] Bellavista P., Corradi A., Stefanelli C., An Integrated Management Environment for Network Resources and Services. IEEE Journal on Selected Areas in Communcations, Vol. 18, No. 5, May 2000
[3] Boutaba R., Polyrakis A., COPS-PR with Meta-Policy Support, IETF Internet Draft, May 2001.
[4] Boutaba R., Polyrakis A., Projecting Advanced Enterprise Network and Service Management to Active Networks, IEEE Network, Jan./Feb. 2002
[5] Bredin J., Kotz D., Rus D., Economic Markets as a Means of Open Mobile-Agent Systems, In the workshop "Mobile Agents in the Context of Competition and Cooperation" at Autonomous Agents, May 1999
[6] Bredin J., Maheswaran R. T., Imer C., Basar T., Kotz D., Rus D., A Game-Theoretic Formulation of Multi-Agent Resource Allocation, 2000
[7] Brunner M., Active Networks and its Management, NEC-NDLE-IR-2001-5, Feb. 2001
[8] Casassa M., Baldwin A., Goh C., POWER Prototype: Towards Integrated Policy-Based Management, IEEE/IFIP Network Operations and Management Symposium, 2000.
[9] Chan K., Durham D., Gai S., Herzog S., McCloghrie K., Reichmeyer F., Seligson J., Smith A., Yavatkar R., COPS Usage for Policy Provisioning, IETF Internet Draft, draft-ietf-rap-pr-05.txt, Oct. 2000. [RFC 3084]
[10] Cheikhrouhou M. M., Conti P., Labetoulle J., Intelligent Agents in Network Management: A State-of-the-art, 1998
[11] Dobson J.E., McDermid J.A., A Framework for Expressing Models of Security Policy, IEEE Symposium on Security & Privacy, May 1989, Oakland, CA, 1989.
[12] Goldszmidt G., Yemini Y., Distributed Management by Delegation, Proceedings of the 15th International Conference on Distributed Computing Systems, June 1995
[13] Hegering H., Abeck S., Neumair B., Integrated Management of Network Systems, pg.6, Morgan Kaufmann Publishers, Inc. 1999
[14] Hegering H., Abeck S., Neumair B., Integrated Management of Network Systems, pg.82-94, Morgan Kaufmann Publishers, Inc. 1999
[15] Hegering H., Abeck S., Neumair B., Integrated Management of Network Systems, pg.121-152, Morgan Kaufmann Publishers, Inc. 1999
[16] Hegering H., Abeck S., Neumair B., Integrated Management of Network Systems, pg.279-287, Morgan Kaufmann Publishers, Inc. 1999
[17] Hu C., Chen W. E., A Mobile Agent-Based Active Network Architecture, ICPADS 2000
[18] IETF Internet Draft: Policy Framework, draft-ietf-policy-framework-00.txt, work in progress, Sept. 1999.
[19] IETF Internet Draft: Policy Terminology, draft-ietf-policy-terminology-00.txt, work in progress, July 2000.
[20] IETF RFC 2748: The COPS (Common Open Policy Service) Protocol, IETF RFC 2748, Jan. 2000.

[21] Ju H., Choi M., Hong J., EWS-Based Management Application Interface and Integration Mechanisms for Web-Based Element Management, Journal of Network and Systems Management, Vol.9, No.1, 2001
[22] Knight G., Hazemi R., Mobile Agent-Based Management in the INSERT Project, Journal on Network and Systems Management, Vol.7, 1999
[23] Koch T., Kramer B., Rohde G., On a Rule Based Management Architecture, The 2nd International Workshop on Services in Distributed and Networked Environments, IEEE Computer Society, Whistler, Canada, 1995.
[24] Koch F. L., Westphall C. B., Decentralized Network Management Using Distributed Artificial Intelligence, Journal of Network and Systems Management, Vol.9, No.4, Dec. 2001
[25] Lange, D., Java Aglets Application Programming Interface (J-AAPI), IBM white paper, Feb. 1997 (www.trl.ibm.com/aglets/JAAPI-whitepaper.htm)
[26] Liotta A., Pavlou G., Knight G., A Self-Adaptable Agent System for Efficient Information Gathering, 2001
[27] Lupu E., Sloman M., Conflicts in Policy-Based Distributed Systems Management, IEEE Transactions on Software Engineering, Vol. 25, No. 6, Nov. 1999.
[28] Martin-Flatin J., Push vs. Pull in Web-Based Network Management, Technical Report SSC/1998/002, Swiss Federal Institute of Technology Lausanne, 1998
[29] Martin-Flatin J. P., Znaty S., Hubaux J. P., A Survey of Distributed Enterprise Network and Systems Management Pradigms, Journal of Network and Systems Management, Vol.7, No.1, 1999
[30] Moffett J., Sloman M., Policy Hierarchies for Distributed Systems Management, IEEE Journal on Selected Areas in Communication, Vol. 11, No. 9, Dec. 1993.
[31] Murphy S., Lewis E., Puga R., Watson R., Yee R., Strong Security for Active Networks, IEEE OPENARCH 2001
[32] OMG, The Common Object Request Broker: Architecture and Specification, v2.3, Jun. 1999
[33] Prozeller P., TINA and the Software Infrastructure of the Telecom Network of the Future, Journal on Network and System Management, Vol.5, Dec. 1997
[34] Rogerson D., Inside COM, Redmond, WA, Microsoft, 1997
[35] Straber M., Baumann J., Fohl F., Mole – A Java Based Mobile Agent System, 10th European Conference on Object-Oriented Programming ECOOP'96. Jul. 1996
[36] Tennenhouse D. L., Smith J. M., Sincoskie W. D., Wetherall D. J., Minden G. J., A Survey of Active Network Research, IEEE Communications Magazine, Vol. 35, No. 1, Jan. 1997
[37] Thompson J., Web-based Enterprise Management Architecture, IEEE Communications Magazine, Mar. 1998
[38] Waldbusser S., Remote Network Monitoring Management Information Base. RFC 1757, Feb. 1995
[39] Wooldridge M., Jennings N. R., Intelligent Agents: Theory and Practice, The Knowledge Engineering Review. Vol. 10, No.2, 1995

State of the Art of Service Creation Technologies in IP and Mobile Environments

Jorma Jormakka
Helsinki University of Technology
Networking Laboratory
E-mail: jormakka@tct.hut.fi

Henryka Jormakka
Technical Research Centre of Finland
Information Technology
E-mail: henryka.jormakka@vtt.fi

Abstract: The paper gives an overview of service creation technologies for IP, PSTN-IP interworking and for mobile networks. There is a large number of possible future service creation technologies around, such as Parlay, MExE, XML, JAIN, and the present service technologies, like IN and CAMEL are not likely to disappear. The paper classifies the technologies into four groups and discusses the ideas and limitations of technologies for value added services.

Key words: Service creation, IN, CAMEL, OSA, Parlay, TSAS, VHE, JAIN, CPL, MExE, WAP, J2ME, SAT, PINT, SPIRITS, Mobile agents/Active networks.

1. INTRODUCTION

Next Generation Network (NGN) service technologies form a loosely defined collection of technologies for future service creation. Terms like All-IP, Softswitch, Extensible Markup Language (XML), are implied.

Technologies, which remind of the Intelligent Networks (IN), like Customized Applications for Mobile network Enhanced Logic (CAMEL), or Open Service Access (OSA)/Parlay, are also included. Additionally, Universal Mobile Telecommunications System (UMTS) has its role in the new service technologies. In fact, without services UMTS would most likely fail, therefore service technologies are crucial for the concept. UMTS and 4th generation mobile networks contribute to the area: wireless terminals contain service technologies like Mobile Execution Environment (MExE) and SIM Application Toolkit (SAT). ExE includes Wireless Application Protocol (WAP) and Java 2 Micro Edition (J2ME), which are relevant on their own as well. Also the Virtual Home Environment (VHE) concept has an important role in service creation. VHE is an elaboration of the personal mobility idea, first introduced in the IN as the UPT service.

NGN service technologies can be divided into four groups. The first group follows the Intelligent Network concept. Intelligent Network is today the predominant service creation technology in voice services in GSM and PSTN. To this group belong CAMEL - an IN technology which allows roaming GSM users to access IN services, PINT (PSTN and Internet Interworking) and SPIRITS (Services in the PSTN/IN Requesting Internet Services).

The second group of service technologies has borrowed ideas from IN and TINA (Telecommunications Information Networking Architecture), but do not need IN type Service Control Points (SCPs) for accessing services. This set includes Parlay, OSA, TSAS (Telecommunications Service Access and Subscription) and JAIN (Java APIs for Integrated Networks). Usually these technologies are understood as service technologies for IP networks, but they do not assume that the network is IP.

The third group of service technologies contains technologies, which mostly reside in a wireless terminal. WAP and I-Mode were the first technologies of this type. To this group belongs also J2ME, an edition of Java running in small wireless terminals like mobile phones, MExE - an enhancement of WAP and J2ME, and SAT - a related technology, where service logic resides in the SIM or Universal Subscriber Identity Module (USIM) card.

The fourth group is a loosely defined set of additional service technologies. Active networks and Mobile Agent technologies fall into this group.

In the presented paper the most important service creation technologies and their current state are described. The conclusions discuss the limitations and future of service creation.

2. SERVICE CREATION TODAY

Currently service creation in the circuit-switched side and in IP networks is based on different technologies.

- *Service creation in PSTN/IN:* The services consist of supplementary services of ISDN made as additions to the protocol, and of IN services. New services are created as IN services using Service Creation Environments (SCEs), which enable definition of services as Service Independent Building block (SIB) chains. Trigger tables must be filled for giving triggering rules for Trigger Detection Points (TDPs), but for creation of new simple services this is enough. Totally new service types require implementation of new Capability Sets (CSs) of IN. Currently only IN CS1 and parts of IN CS2 are in commercial use.
- *Service creation in the Internet:* Basic Internet data services, telnet, ftp, smtp, have been created as data communication protocols on the application layer. Protocol development meant writing Unix-style programs on top of the socket interface, as no specific service creation methods have been used. Only after the invention of HTTP, there exists now a service creation method: WWW services are created using HTML, Java applets, Common Gateway Interface (CGI), or e.g. Perl scripts.
- *Service creation in 2G:* The early way of creating services was adding GSM supplementary services to the protocol. Many new services can be created using these inbuilt features. IN technology CAMEL is needed for making services that roaming users can use, but CAMEL is also seen as a flexible way to create new services. GSM-based 2G services include: Voice calls with GSM supplementary services, group calls etc., Packet data call, Short Message Service (SMS), WAP services: games, entertainment, ticket booking etc. On top of these services is the vendor specific IN platform with CAMEL.

As a summary, in all the three areas service creation has been separated from protocol development. There is a need for a service platform and a service creation process by which new services can be created fast. A natural short-term goal has been to integrate the good sides of service creation methods enabling PSTN/IN type services in IP networks and IP type services in the PSTN/mobile network side. A long-term goal is to create a safe, easy to use, open service creation environment enabling new business opportunities, that is, creation of the service network architecture.

3. FUTURE OF SERVICE CREATION

3.1 IN (Intelligent Networks)

IN [1] service creation follows the IN Conceptual Model (INCM). Service is defined in the service plane (SP) using free form text. A new service should preferably be created using existing service features as otherwise it may be difficult to implement it on a given CS. The purpose of a CS idea is to list targeted services so that it is possible to understand if a given IN platform is sufficient for implementing a new service. In IP service creation, the idea of a capability set has not got enough attention. A probable reason is that giving target services is considered to be similar to defining the services, but that is not the case.

The next step in INCM is to describe the service in the Global Functional Plane (GFP) using SIBs (Service Independent Building blocks). An IN service is a chain of SIBs starting from a Point of Initialization in the Basic Call Process SIB and ending to some Point of Return. In IN CS1 it is not allowed to have parallel SIB chains. In CS2 this is possible and Points of Synchronization are used to synchronize parallel chains.

In the Distributed Functional Plane (DFP) the service is implemented using Information Flows and Functional Entities. In creation of IN services DFP is needed in arming Detection Points. Statically armed Detection Points are called Trigger Detection Points (TDP) while the dynamically charged are called Event Detection Points. Triggering IN service logic occurs in a TDP according to a trigger table. The structure of a trigger table is not standardized, but typically triggering can be based on prefixes or individual numbers. IN dialog with Service Switching Function (SSF) and Service Control Function (SCF) is started at some TDP and the trigger table must be set.

The last plane in INCM is the Physical Plane (PhP). The issues of PhP, which appear in IN service creation, are possible changes to INAP and placement of the Functional Entities. Although INAP is standardized, usually ETSI CORE INAP is used, there are often small changes introduced by operators and a new service may require some additional information to be carried by INAP. Furthermore, the exact signaling sequences can be made in many ways for the same service. Functional Entities naturally are placed in an IN platform to some Physical Entities, but depending on the load, it may be desirable to realize a new service with a different placement of FEs. For instance, one may use a SCP with SCF and Service Data Function (SDF), but SDF may also be placed on a stand-alone Service Data Point. Lastly, there may be some cases when the IN standard is intentionally not followed because of traffic considerations, for instance, in televoting signaling should go to SCF for each call, but in an implementation signaling may go to SCP only once per several calls.

This shows that service creation in the IN mostly occurs in the GFP using SIBs. Special Service Creation Environments (SCE) with graphical user interfaces are applied. Often the SIBs SCEs have do not directly correspond to the standard SIBs. For instance Siemens's EWSD SCE has much more and more specific SIBs than the ITU-T IN Recommendations mention.

The capability sets of IN in use today are IN CS1 and partially IN CS2. IN CS3 is fully standardized and IN CS4 is on draft stage. IN CS4 is leveraged to future networks. It is the most interesting for service creation in IP and mobile networks since it contains PSTN - IP interworking. IN CS4 defines interfaces to Parlay and to PINT. SPIRITS is not mentioned in ITU-T IN CS4, but this technology fits in quite naturally.

3.2 CAMEL (Customized Applications for Mobile network Enhanced Logic)

CAMEL, specified by ETSI/3GPP [13], is a technology of GSM Phase 2+, which enables roaming GSM users to access IN service in their home operator's environment. CAMEL adds some new functional entities to the GSM network.

gsmSSF: Corresponds to IN Service Switching Function for GSM voice calls. The gsmSSF functionality is located to MSC.

gprsSSF: Like gsmSSF but for SMS. This functionality is located to SGSN.

ipSSF: A new SSF functionality will be introduced in CAMEL for IP Multimedia Core Domain in 2GPP R5. ipSSF will be placed in CPS. CPS has interfaces with GCSN, HLR and gsmSCF.

gsmSC:. Corresponds to SCF and SCP, but can communicate with HLR using AnyTimeInterrogation. The functionality resides in a physical point, also called gsmSCF.

CAMEL uses a modification of IN INAP protocol, called CAP (Camel Application Part). It is a protocol, by which gsmSSF communicates with gsmSCF. Additionally CAMEL contains modifications to GSM MAP: a new operation, AnyTimeInterrogation is added to MAP and gsmSCF can ask HLR information of the roaming user.

There are two state machines in gsmSSF: originating and terminating state machines (O-BCSM, T-BCSM). They resemble IN CS1 BCSM, but are simpler. They contain a restricted number of Detection Points, especially CAMEL Phase 1. In later phases of CAMEL more Detection Points have been added to O-BCSM and T-BCSM.

CAMEL Phases 1 and 2 are used, but they only deal with circuit-switched networks. CAMEL Phase 3 is a new addition. It includes data services for General Packet Radio Service (GPRS), like the GPRS prepaid service.

Service creation in 3G is based on VHE scenario including CAMEL, OSA, and MExE. Most UMTS calls will have CAMEL processing. Prepaid service is the main service of CAMEL. It continues to be the main service also in CAMEL Phase 3, this time for prepaid charging of IP traffic.

3.3 Parlay

Opening IN to 3pty service providers has been too risky because of security and reliability considerations. Founded in 1998 the Parlay Group [6] is claiming to have a solution in the form of Parlay APIs, through which 3pty applications can in a controlled and secure way make use of network functionality and become independent of the underlying network technology. The network under the Parlay APIs can be IP, Quality of Service IP (QoS IP), or a circuit switched network.

The Parlay APIs consists of two categories of interface – Framework Interfaces and Service Interfaces. The Framework Interfaces provide applications with basic mechanisms for making use of the service capabilities in the network. An incoming call to a Parlay service comes through the Parlay Framework that offers capabilities necessary for the Service Interfaces to be secure and manageable. The client applications are using it to conduct the authentication process with the Framework provider, select an instance of network's service, or subscribe to new services. Examples of Framework APIs are Authentication, Discovery or Trust Management APIs. After the authentication process the applications use Discovery API in order to find handles to the services offered by the Parlay. The network services are accessible through Service Interfaces that offer applications access to the network capabilities such as call management or user interaction.

Service creation using Parlay in practice means writing objects using CORBA. Typical languages are Java and C++. These objects call Parlay APIs directly. In this way Parlay service creation is similar to application development, but writing Java is considered relatively easy.

There is a newer approach for using Parlay APIs, so called Parlay/X. In this variant, a service is implemented with XML PDUs on top of HTTP using the SOAP-mapping. The XML documents call scripts in any standard XML method (like XPath/XLib). These scripts call Parlay APIs. This service creation method is similar to CPL, or HTML/CGI service creation.

3.4 OSA (Open Service Accesss)

OSA, the standardization effort of 3GPP [14], specifies architecture that enables applications to make use of network functionality through open standardized interfaces. It resembles the Parlay architecture and can be considered as a specialization of Parlay to mobile networks. OSA consists of three parts:

Applications implemented in application servers,

Framework, which as in case of Parlay provides applications with basic mechanisms enabling them to use service capabilities in the network. The mechanisms are provided by framework capability features such as Authentication, Discovery, or Service Agreement,

Service Capability Servers providing applications with service capability features such as Call Control or User Location

The Framework's and Service Capability Servers' Service capability Features (SCF) are provided through framework and network interfaces respectively.

Currently the standardization bodies of 3GPP and Parlay Group work to achieve a common OSA model and aim to synchronize the release dates. However, although they specify the same functionalities, the terms are still different. The current API is known as OSA Release 5 or Parlay 3.0.

3.5 TSAS (Telecommunications Services Access and Subscription)

Influenced by the TINA business model and service architecture, the OMG Telecommunications Domain Task Force issued in June 2000 document called "Telecommunications Service Access and Subscription" in which it specified interfaces for handling business transactions among three parties: service providers, retailers and consumers. One of the objectives of TSAS was to improve scalability of the TINA specification and remove the redundancy between TINA objects' interfaces using segmentation. Originated from TINA access session, TSAS model provides segmentation of interfaces related to access and subscription.

The document describes how services can be retailed on behalf of service providers, which in turn offer their services to the retailers. It concentrates on specifying consumer-retailer and retailer-service provider interfaces. The interfaces enable users to subscribe, select and access services that are customized according to user preferences. The subscription interfaces and operations have been reused in Parlay. The TSAS specification is technically aligned with these of the Parlay Group, which means that the service interfaces specified in the Parlay API can be offered using this specification.

3.6 VHE (Virtual Home Environment)

VHE is a concept of the 3G mobile system for personal service environment (PSE) portability across network boundaries and between terminals which means that the user will have the same interface and service environment regardless of location. PSE is a combination of a list of subscribed services, service preferences and terminal interface preferences. It also encompasses the user management of multiple subscriptions, e.g. business and private, multiple terminal types and location preferences. The PSE is defined in terms of one or more User Profiles.

The VHE concept originated from the 3GPP specification [12], which specifies the VHE business model and requirements for a possible service architecture realizing the concept. The work on VHE conducted by 3GPP and other fora is for promoting a service architecture providing an open Application Programming Interface (API) that resides between the application layer and the service component layer. The API offers access to network information and enables services operated by enterprises outside of the network domain to access the network capabilities. The VHE concept can be realized using existing toolkits such as OSA, Parlay, MExE, CAMEL or USAT.

3.7 MExE (Mobile Execution Environment)

MExE [5] is a specification introduced by ETSI and developed by 3GPP enabling provision of standardized execution environment in user equipment (UE). The UE consists of the Mobile Equipment (ME) and SIM/USIM (Universal Subscriber Identity Module). MExE provides the ability to negotiate the UE supported capabilities with a MExE service provider, in this way allowing applications to be developed independently of any UE platform or operator's service platform. The applications can be executed on a remote server, or downloaded to the UE and directly executed there. Two UEs may also be engaged in a MExE service with each other, with the network effectively supplying the "pipe" and not playing a MExE role in the connection.

To handle the variety of possible UE configurations MExE divides them into so called Classmarks. Currently there are defined three Classmarks with more to follow in the future. Classmark 1 terminals are based on WAP, Classmark 2 groups PersonalJava enabled devices, Classmark 3 terminals are based on the J2ME Connected Limited Device Configuration (J2ME CLDC).

The technologies in MExE, WAP, PersonalJava and J2ME CDLC MIDP, are important without MExE. The main additions of MExE are in the security and access rights model. Due to the support of ETSI and 3GPP, it is likely that MExE will be supported by vendor equipment in the near future.

3.8 WAP (Wireless Application Protocol)

WAP is the WAP Forum's protocol for bringing the Internet to mobile phones. In WAP 2.0, two protocol stacks are supported and they are combined on the Wireless Application Environment (WAE) layer [8], [9]. The WAP 1 stack consists of Wireless Session Protocol (WSP), Wireless Transaction Protocol (WTP), Wireless Transport Layer Security (WTLS) and Wireless Datagram Protocol (WDP), while WAP 2 stack contains modified TCP/IP protocols: WP-HTTP, TCP*, TLS, Wireless IP. WAP works on any bearer, like CSD, SMS, USSD, GRRS and on any network, like GSM, UMTS, PDC-P, iDEN, CDPD.

The WAP Forum has defined an XML based language Wireless Markup Language (WML). XHTML (eXtendable HyperText Markup Language) is another possibility in WAP Version 2.0. XHTML is a version of HTML following the XML standard, including ending tags and defining DTD in the header. WML definition contains so called deck with several WML pages. The deck is similar to HTML front page, but WML is optimized for over the air interface. A deck is moved as whole and pages of the deck will be locally accessible in the mobile phone. Unlike HTML, WML keeps the state, so going to next or previous page is possible in a WML deck, has timers and variables.

WAP services are typically created using a HTTP-server and a WAP simulator. Creating a service means writing a deck of WML pages. It is put to the server and can be accessed by the WAP phone simulator. WMLScript scripts can be added to the WML pages to give some functionality, J2ME MIDlets can add more complex functionality.

Wireless Telephony Application Interface (WTAI) is another way of creating voice services in WAP. It means calling the WTAI API and is mostly intended for authorized WTA servers.

WAP works mostly in the pull-mode, like HTTP. The currently interesting WAP Push of Version 1.2 makes possible new service types.

From performance point of view, WML has optimizations: WAP Binary XML (WBXML, WMLC) and Wireless Bitmap format (WBMP) for images for the Over-The-Air interface. Still, WML is not especially compact. As a language derived from XML, WML carries all tags for elements. This is useful for writing browsers understanding pages obtained from anywhere in the Web, but transmitting tags is a waste of bandwidth. WAP access to the Web is through WAP proxies and probably a language more compact than WML is feasible.

The version of WAP supported by most mobile phones is WAP Version 1.1., released by the WAP Forum in June 1999. The current WAP 2.0 version was released August 2001. Version 1.2.1 was still supported by only few models at the end of 2001. Despite of the slow market development, WAP has a strong support as 99% of mobile phone vendors are in the WAP Forum. WAP is still a promising future technology with high hopes for WAP on GPRS.

3.9 J2ME (Java 2 Micro Edition)

Java 2 Standard Edition (J2SE) does not fit into small mobile devices, like mobile phones. Several smaller size versions of Java have been created, like Personal Java for set-top boxes and Personal Digital Assistants (PDA). J2ME is a stripped-off version intended to a range of small devices from embedded computers to mobile phones.

J2ME has two configurations: Connected Limited Device Configuration (CLDC) and Connected Device Configuration (CDC). CDLC is intended to mobile phones and small PDAs, while CDC is intended to replace Personal Java in set-top boxes and more powerful PDAs. CDC is much more like J2SE while CDLC has rewritten virtual machine and restricted rewritten libraries. On top of a configuration are profiles. Examples of CDC Profiles are Personal Profile and Foundation Profile, on CDLC there are for instance MIDP (Mobile Information Device Profile) and PDA profile. MIDP is of special interest as it will be used in mobile phones and can enhance WAP. In general, J2ME should not be seen as a replacement of WAP, but as adding new power.

Java MIDlets correspond to J2SE and J2EE (Enterprise Edition) applets. MIDlets are usually packed to Java JAR-files, which are downloaded over the air from a HTTP server using WAP and executed in the User Equipment. J2ME services are developed using simulators in the same way as WAP services. There are special concerns of size and performance when writing MIDlets, and standard libraries of J2SE usually do not exist. Differences in presentation of MIDlets in implementations of different vendors must often be taken into account with small vendor-specific code segments.

MIDP Over-The-Air (OTA) provisioning partially supports service discovery, installation and upgrading, but retrieval of MIDlets is outside the scope of the current MIDP specification. The mobile equipment must contain a Discovery Application (DA) and using DA new JAD-files can be downloaded from the Web.

Currently, J2ME runs a bit too slowly for good customer satisfaction and it has not yet filled the promise of Java: write once, run anywhere.

3.10 SAT (SIM/USIM Application Toolkit)

The (U)SIM Application Toolkit (SAT/USAT) of 3GPP [11] is designed to provide a standardized execution environment for applications stored on the (U)SIM card. It provides mechanism, which enables the applications to interact with any mobile equipment, which supports the specified mechanism(s) thus ensuring interoperability between a USIM/SIM and the mobile equipment, independent of the respective manufacturers and operators. Additionally, a transport mechanism enabling applications to be downloaded and/or updated is provided. SAT/USAT applications are more limited than MExE applications, but otherwise SAT/USAT can provide nearly the same capabilities as MExE and ensures a larger degree of freedom, because of more secure SIM environment.

SAT/USAT is one interesting element for the provision of VHE. It specifies the interaction between the SIM card, the ME and the network. The standardisation of these interactions contributes to satisfying the portability requirement of the VHE. Indeed, SIM/USIM cards and their applications become portable over several terminals and networks. Additionally, the increased intelligence in the terminal side will cause that users can start managing their services more independently from their operator. This may reduce the need for service provisioning, but also bring provisioning to a new level – inside the mobile terminals.

3.11 PINT (PSTN IP Interworking)

PINT by IETF [2] is a simple way to use IN services from IP networks. The idea of PINT is that an IP host sends a request to a PINT server. The PINT server relays the request to a PSTN network element, typically SCP, but possibly Private Branch Exchange, MSC, or some other element in a circuit-switched network. Finally, the circuit-switched network performs the action. Though PINT uses SIP, it is not connected with IP telephony. Let us give a sample PINT service: an IP host requests using the Internet that an IN SCP causes a fax to be sent to a fax machine using PSTN (Request to Fax Content). PINT is easy to implement, but it is often considered as a transitional solution as Parlay makes similar things without the IN SCP.

The PINT protocol is developed from Session Initialization Protocol (SIP) and Service Description Protocol (SDP). SIP can be considered known. SDP is a protocol for multimedia session description, especially for joining in multimedia sessions.

Initially, the following PINT services are defined:

- Request to Call
- Request to Fax Content
- Request to Speak/Send/Play Content

3.12 SPIRITS (Services in the PSTN/IP Requesting Internet Services)

SPIRITS [3] is a new SIP-based protocol by IETF, currently on draft stage. SIP here is not used for establishment of an IP call, and in fact, like PINT, SPIRITS does not have anything to do with IP telephony. SIP is here used as a general purpose signaling protocol and the goal of SPIRITS is to allow PSTN users to access through SCP services in the IP side, and to allow IP users to access IN-based services in the PSTN.

Some services can be made with either SPIRITS or PINT: for instance click-to-dial from a Web page can be a PINT service, or it can be a SPIRITS service. It is also possible to make this service using Parlay. In general, there is nothing in SPIRITS that could not be made with CPL and Parlay, but the main goal of SPIRITS is not to add services that could not be made with other technologies. SPIRIT is mainly for leveraging the present IN platforms so that some IP services can use the present IN. This is important especially in countries where IN is new and the investments are not yet paid off. Therefore there is a strong interest in using the IN platforms as long as they have been planned to be in operation.

The motivation for defining SPIRITS was to provide the following services:

- Internet Call Waiting,
- Internet Caller-ID Delivery,
- Internet Call Forwarding.

3.13 CPL (Call Processing Language)

CPL is one of the XML-based languages created for value added services. Using CPL it is possible to implement simple telephony services, like forwarding and redirecting calls. Voice XML (VXML) is another similar scripting language, for voice controlled services. CPL is defined by IETF and used in SIP-based IP telephony but it is not tied to any particular protocol. In general, SIP uses scripting languages for services in which it differs from e.g. H.323 which has supplementary services built into the protocol (H.450).

Any XML based scripting language is defined by a Document Type Definition (DTD), which defines elements and attributes. The elements then correspond to states in the script or to methods of some API. CPL like any XML language has a tree structure. It contains on the top level tags *ancillary*, *subaction*, *outgoing* and *incoming*. The actual script is under the last three tags. CPL script contains the following: *switches*, *location modifiers*, *signaling operations*, and *non-signaling operations*. A typical CPL script could for instance use *address* switch to make a decision based on the address, it could retrieve the target location using the location modifier *lookout*, then it could use the signaling operation *proxy* to forward the request to the given location. Additionally the script should handle all error cases, like if the receiver is busy, there is no answer and so on.

XML suits rather well to service creation. However, XML is basically a browser language. It has powerful additions for browser usage, like XSLT, XPath and Xlink. An XML document carries starting and ending tags as ASCII names, also data is in ASCII. Any compression should be on lower layers. For high bandwidth connections this is fine and it helps a browser to display unknown data. However, in XML based data transported over the air interface there is less sense in carrying all tags as full names: both sides of communication must in any case know the methods of the API to be called. In the favor of XML one must admit that features of XML as a browser language are useful in graphical representation and editing of XML scripts.

SIP-CGI is a more powerful scripting language for IP telephony. SIP-CGI scripts can execute arbitrary programs in a SIP server. CGI causes some problems because it is so powerful: poorly used HTTP-CGI has caused serious security problems, like the Windows'2000 IIS bugs.

J2EE Servlets from suitable Java beans package are another possibility, and in this case Java security mechanisms apply. In IP telephony there are proposals for SIP Servlets. In expression power SIP Servlets are similar to SIP-CGI.

3.14 JAIN (Java APIs for Integrated Networks)

The objective of JAIN Community initiative was to create a number of integrated network APIs [10] that abstract the details of networks and protocols implementations and provide service portability, convergence, and secure access to the networks. JAIN integrates service creation in wireless, packed based and wireline networks by separating service-based logic from network-based logic what can be considered as separation into application and protocol layers respectively.

The protocol layer of JAIN is based on Java standardization of specific protocols like SIP, H.323, MAP, etc. It provides the JAIN protocol API that comprises Java APIs for the protocols. As a result applications and protocols stacks can be dynamically interchanged, what can help to avoid current situation when introduction of a new variant of protocol requires upgrading of application-level software.

The application layer provides a single call model JTAPI (Java Telephony API) across all protocols supported in the protocol layer. It is accessible through the Application API that consists of the JAIN Call Control (JCC) API, JAIN Coordination and Transaction (JCAT), JAIN Service Logic Execution Environment (JSLEE) API and JAIN Service Provider APIs (SPA). The two first types of APIs provide interfaces to generic call model supporting access to advanced intelligent network. JSLEE API provides access to a consistent run-time environment and a common model for deploying services in Java, while SPAs provide a Java version of Parlay API.

Service development in JAIN depends on the level of trust the operator of the JAIN environment has on the service provider. Trusted service providers create services using Java servlets and XML, typically Java servlets are given by some Java beans packet. Untrusted service providers access the JAIN environment through Parlay APIs.

JAIN does not build on CORBA, as was the case with Parlay, but on a Java platform: telephony is realized with JTAPI, and different networks can be used under the Java platform. The XML-based technologies, enhanced with Java servlets, can be used for service provisioning for trusted service providers, while JAIN-Parlay APIs are used for untrusted 3pty service providers.

3.15 Some other ideas

TINA: TINA [7] is already an old effort to create a service platform on CORBA realizing the ODP (Open Distributed Processing) model. Though TINA ideas are still continued in the OMG, the architecture's main contribution to future service creation may well be on the level of reuse of ideas. TINA's auxiliary projects, such as Dolmen made successful demonstrations of certain enhancements of CORBA mobility and they may be reflected in CORBA.

Mobile agents: The use of mobile agents (software programs capable of moving in a computer network and acting autonomously on behalf of some entity, however under the control of some authority) can be regarded as a powerful asset for implementing complex and highly dependable distributed software systems. Agents, in fact, may help to better exploit the TINA type paradigm. This is because the main characteristic of agents, that is to play in autonomy in visited environments with reference to policies set by the originating environment although in respect of constraints imposed by the host nodes, seem naturally suited to model the interactions between components belonging to different business domains. Otherwise stated, mobile agents, in particular, can take along the service logic as well as the user profiles when migrating between different network nodes. It is expected that service availability and flexibility will be considerably increased, while signaling traffic load may be kept to controllable levels. Of course, the price to pay is the availability of an agent execution environment, an "agency", in all network and user nodes. Additionally, problems of scalability, reliability, security and quality of service have to be faced.

Active networks: Active networks are of two types, active packets and mobile agents. The latter are already explained. Active packets are small programs programming routers or other network nodes. Security concerns of such a proposal are considerable. Active packets are not intended for the type of value added services discussed here. They have some potential, for instance in protection against Distributed Denial-of-Service attacks. Ability to program packet handling mechanisms is useful for services, however, the overhead required to make such a service platform secure may imply packets that are too long.

4. DISCUSSION

There are many technologies for creation of future services and it is easy to be lost in details. Cost, maturity, supported functionality and other criteria are often used for comparison of technologies. This kind of comparison suits poorly to future technologies, which are on an early development stage. We compare the alternatives briefly using three methods: finding the good ideas of the technology, deducing what limits and requirements the set of technologies has, and classifying the technologies into groups.

4.1 Finding Good Ideas

In this method the important part is the good ideas of the considered technology. There are not that many good ideas behind the service creation technologies. Detecting the promising ideas and reasons why the technologies implementing these ideas have succeeded or failed is one way of predicting the future development. That is, good ideas, which are ignored, rise again in another disguise, and bad ideas tend to end up with similar problems every time they are tried.

Many good ideas have originated in the IN, such as:

- Separation of service logic from basic connection control, and routing to separate service points simplifies introduction of new services.
- External service providers can create services. Although IN did not enable external service providers to offer services on their own platforms but on the operator's platforms, still the idea that there are more roles than operator-customer derives from the IN.
- Services are easier to develop in scripting languages (SIBs in the IN) instead of programming languages as in protocol development. Notice that in TINA this separation was not made. TINA is not used. We see also a similar tendency towards separation of service creation and protocol development in creation of Parlay/X.
- Service creation and provisioning must be a fast process in a competitive situation.
- Capability sets and targeted services are a way to develop service creation platform in phases.
- Personal mobility in the form of UPT is a promising idea.

CAMEL has basically two good ideas:

- Not to reinvent a wheel, IN is a good service creation technology.
- Successful IN services can be made by alternative charging (prepaid).

Some ideas derive from WWW:

- Use of markup languages for service creation (HTML, WML, XML).
- Use of scripts (CGI, CPL), Java applets, Java servlets to enhance the services.

Very importantly, the Internet and WWW showed that successful end-system data services are possible, but here we concentrate on value added services using network resources.

A few ideas were originated in ODP and TINA:

- Use of CORBA platforms, continued by Parlay.
- Dynamic service discovery.
- Importance of the business models as an elaboration of the IN three roles (operator, subscriber, and user). The business model comes from the enterprise viewpoint of the ODP design process and is included in UML

as use cases. However, the business aspect of the business model is a very important and often understudied concept for new services.

Parlay and OSA have elaborated the idea of an external service provider, already clearly defined in TINA:
- External service providers can be allowed to access network resources via secure APIs.

WAP, MExE, J2ME have one good idea:

- Mobile networks are a success, the Internet is a success, let us merge them and it should be a success.

VHE has elaborated the idea of Personal Mobility to:

- User should be able to use all services as if he was in his home environment.
- Single-sign-on can be understood as an idea related to the Personal Mobility and VHE.

The idea in Mobile agents is autonomous agents forming an agent community where the members are discussing with each other in agent languages, This AI idea has not been realized and the technology is not in the among the most probable service creation methods, mostly due to security considerations. Active packets of active networks have even more serious security concerns than mobile agents.

There are also some doubtful claims about services in which many strongly believe:

- There are many useful services to be invented that users will pay for. There actually are some successful services: IN FreePhone, PremiumRate, VPN, Televoting, CAMEL Prepaid, Internet WWW, email, file transfer, but most services find few users.
- There are some ideas that sound good: brokering of services as a business idea, eCommerce, mCommerce, VHE. The doubtful claim is that they are good enough.

4.2 Limitations And Requirements

End system data services, like the Web, are proven. Many of the service creation technologies try to realize commercially successful value added services, which use network resources. This idea used to be alien to the

Internet. There are limitations arising from this goal. Value added services using network resources are mostly build on the following possibilities. Changing routing (examples: IN VPN), modifying charging (examples: IN FP, PR), avoiding assumptions in network dimensioning (example: televoting), providing anonymity to the receiver or to the caller (examples: IN PR), adding specialized resources (examples: answering machines, DTMF), locating users (examples: Camel locate in Parlay), finding information (examples: WWW search machines, brokering). These enable a large range of services, but not necessary to the extent that is often thought. Most new good value added service ideas are economically failures when they are tried.

As for requirements, there are some coming from the goals of service creation. For competitive reasons services should be created by external service providers in a very short time. This means that the service creator cannot be expected to write a secure service. It must be the service platform that takes care of security, management, charging etc. supporting activities. This is a positive shift from the Internet style of protocol development directly on top of the socket interface on the weakly typed C-language using Unix system calls running as root, having given us all root exploit problems in the past. If the service platform is secure enough, maybe the bug problem can be solved. However, the mechanisms to make a secure platform, like sandboxes and access rights given to authenticated code, are not sufficient as they restrict the services that are possible. Special attention should be paid to standardizing the requirements of a service platform and to specification of new tools which can check that the requirements are filled. Security and reliability are one part of the requirements.

4.3 Classification

Main stream future service creation technologies seem to fall into three groups: IN-based services, Parlay-based services and MExE-based services. These groups are likely to coexist in the future hybrid network and they can be seen as complementary.

The IN-based technologies include IN, CAMEL and PINT/SPIRITS. IN is a mature service creation technology and CAMEL will very likely be heavily used in 3G. In this scenario, the IN is in the PSTN side and seems like a technology that will disappear, but one should not forget IN CS4. IN is actually developing towards a service creation technology for hybrid networks.

The services in the operator's side in all-IP networks are likely to use Parlay APIs. The technologies Parlay, OSA and TSAS all might converge to

Parlay. VHE is a future goal of these technologies and it may mean a great change in the service field. For the time being, VHE is still not mature enough. JAIN can also be included in this group, though JAIN is considered as a technology for hybrid networks. JAIN and Parlay/X combine XML-based service creation to this set of technologies. We should expect to see Parlay API, Java servlets and XML-based service languages in the IP network.

The last set of service creation technologies concentrates on services to mobile equipment. WAP and J2ME may be deployed alone, but in 3GPP they are combined under the MExE technology. MExE actually adds only some security features to these technologies, but it is likely that MExE will be used. There is no competition between the MExE set of technologies and the Parlay set of technologies. Because of CAMEL and PINT/SPIRITS there is also no conflict with MExE and IN-based technologies. There may be some competition with Parlay type ideas and IN/PINT/SPIRITS, thought the first suits to All-IP better and the latter is essentially a solution to hybrid networks.

These three sets of service technologies can bring a mobile, IP or PSTN user value added services which resemble the services offered by IN and WWW. The role of the remaining technologies is to go further: IN and WWW are still very limited in their services. Ideas, such as brokering, trading, auction, intruder response, management and so on will require new technologies. This is where mobile agents, active networks, old TINA ideas and other technologies that seem to be off the main stream come to play a role.

This paper has presented very briefly the main service creation technologies and illustrated some approaches for their comparison in an early stage. Service creation is only one aspect of service development, there are also service provisioning and service management solutions. Due to the limited space and time we cannot discuss these technologies, but they should be considered when comparing the service technology alternatives.

Acknowledgement

The authors want to thank their colleagues and students for fruitful discussions that made the paper possible.

5. REFERENCES

[1] I. Faynberg at al, "The Intelligent Network Standards: Their Application to Services", McGraw-Hill, 1997
[2] S. Petrack, L. Conroy, "The PINT Service Protocol: Extension to SIP and SDP for IP Access to Telephone Call Services, Proposed Standard", RFC 2848, June 2000
[3] L.Slutsman, at al, "The Spirits Architecture", draft-ietf-spirits-architecture-03.txt, February 2001
[4] ITU-T, Recommendation Q.1241, "Introduction to Intelligent Network Capability Set-4", 07/01
[5] MExE Forum website http://www.mexeforum.org
[6] Parlay Group WWW-home page http://www.parlay.org
[7] TINA-C website http://www.tinac.com
[8] WAP-169-WTA-20000707-a. Wireless Telephony Application Specification.
[9] WAP-190-WAESpec-20000329-a. Wireless Application Environment Specification.
[10] website http://java.sun.com/products/jain/
[11] 3G TS 22. 038 "USIM/SIM Application Toolkit (USAT/SAT); Service description"
[12] 3G TS 22.121 "The Virtual Home Environment"
[13] 3GPP TS 23.078, "Customized Applications for Mobile network Enhanced Logic (CAMEL)", March 2001
[14] 3GPP TS 23.127 "Virtual Home Environment/Open Service Architecture", June 2001.

QoS, Security, and Mobility Management for Fixed and Wireless Networks under Policy-based Techniques

Guy Pujolle and Hakima Chaouchi
LIP6, University of Paris 6, 8 rue du Capitaine Scott, 75015, Paris, France

Abstract: This paper introduces a general policy-based management framework within an IP network. It describes how a policy-based approach can be applied to deal with QoS, security, access control, mobility, etc. The framework presented here is derived from IETF works in Policy Framework working group and in Resource Allocation Protocol working group. Then, we describe some applications that could be handled by policy-based systems. Finally, we present some new evolutions that could be part of the future global policy-based networking architecture.

Key words: Internet, QoS, Security, Mobility, Policy-based Management

1. INTRODUCTION

The policy-based networking concepts are born from the need to get an overall end-to-end strategy to correlate the business with the overall network actions. Policy-based networking objectives are to deliver a comprehensive architecture that allows the merging of users, applications and resource policy information with network policy actions. The goals of policy-based networking architecture are to address the enforcement of policies in the nodes of the network and to globally manage the system.

A policy may be defined following two perspectives: an explicit goal and actions to guide and determine present and future decisions. Policies are a set of rules to control and manage network resources.

A policy-based networking system defines two main components: a policy decision point (PDP) and policy enforcement points (PEP). The signaling protocol COPS (Common Open Policy Service) is used to communicate policy information between policy enforcement points (PEP) and a remote policy decision point (PDP) within the context of a particular type of client. To get local policy decisions in the absence of a PDP, the PEP can use the optional local policy decision point (LPDP).

This architecture is described in different RFCs [1-9] mainly coming from the work of the rap (Resource Allocation Protocol) and policy (Policy Framework) working groups.

In this paper, we present policy-based networking operations and some new ideas that could be developed to reach a homogeneous structure to control future IP networks. In section 2, we describe the classical policy-based networking (PBN) architecture. In section 3, we present some applications that could be handled by policy-based systems. In section 4, some specific extensions that we developed in our laboratory are introduced. Finally we present some concluding remarks.

2. THE BASIC ARCHITECTURE

A policy-based networking (PBN) system needs several components:
- a policy management tool,
- a policy repository,
- a policy decision point (PDP),
- policy enforcement points (PEP).

These components are shown in Figure 1. The policy management tool assists the network manager in the task of constructing and deploying policies, and monitoring status of the policy-managed environment. The policy management tool may be seen as an interface between the network manager and the policy repository.

The policy repository can be defined from two perspectives. First, it can be a specific data store that holds policy rules, their conditions and actions, and related policy data. A database or directory would be an example of such a store. Second, the policy repository may be seen as a logical container representing the administrative scope and naming of policy rules, their conditions and actions, and related policy data. A QoS policy, a security policy or a mobility domain would be an example of such a container.

The policy decision point (PDP) is a logical entity that produces policy decisions for itself or for other network elements that request such decisions. A decision involves actions for enforcement when the conditions of a policy rule are true. Policy enforcement points (PEP) are logical entities that enforce policy decisions.

The PEP may also have the capability to select a local policy decision via its local policy decision point (LPDP). However, the PDP remains the authoritative decision point at all times. This means that the relevant local decision information must be relayed to the PDP. That is, the PDP must be granted access to all relevant information to select a final policy decision. To facilitate this functionality, the PEP must send its local decision information (using its LPDP) to the remote PDP.

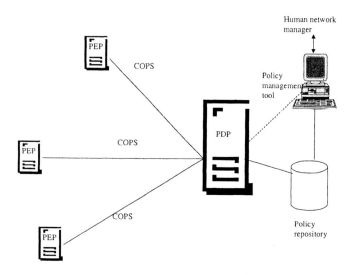

Figure 1 – The basic PBN architecture

The signaling protocol, COPS, is a simple query/response TCP-based protocol that can be used to exchange policy information between a PDP and its clients, the PEPs. Then, a PEP is responsible for initiating a persistent TCP connection to a PDP. The PEP uses this TCP connection to send requests to and receive decisions from the remote PDP. Communication between the PEP and remote PDP is mainly in the form of a stateful request/decision exchange, though the remote PDP may occasionally send unsolicited decisions to the PEP to force changes in previously approved request states. The PEP also has the capacity to report to the remote PDP that it has successfully completed performing the PDP's decision locally. This capability is useful for accounting and monitoring purposes. The PEP is responsible for notifying the PDP when a request state has changed on the PEP. Finally, the PEP is responsible for the deletion of any state that is no longer applicable due to events on the client side or decisions issued by the PDP. When the PEP sends a configuration request, it expects the PDP to send configuration data via decision messages as applicable for the configuration request. When a policy is successfully installed on the PEP, the PEP has to send a report message to the PDP confirming the installation. The server may then update or remove the configuration information via a new decision message. When the PDP sends a decision to remove a configuration from the PEP, the PEP will delete the specified configuration and send a report message to the PDP as a confirmation.

COPS protocol is designed to communicate self-identifying objects which contain the data necessary for identifying request states, establishing the context for a request, identifying the type of request, referencing previously installed requests, relaying policy decisions, reporting errors, providing message integrity, and transferring client specific/namespace information.

To distinguish between different kinds of clients, the type of client is identified in each message. Different types of clients may have different client specific data and may require different kinds of policy decisions. It is expected that each new client-type will have a corresponding usage RFC specifying its interaction within COPS protocol.

The COPS context object identifies the type of request and message that triggered a policy event via its message type and request type fields. COPS identifies three types of outsourcing events: (1) the arrival of an incoming message (2) the allocation of local resources, and (3) the forwarding of an outgoing message. Each of these events may require different decisions to be complete. The content of a COPS request/decision message depends on the context. A fourth type of event is useful for types of clients that wish to receive configuration information from the PDP. This allows a PEP to issue a configuration request for a specific named device or module that requires configuration information to be installed.

There are two mechanisms by which resources may be allocated: configured (or provisioned, or pre-defined, or pro-active) and signaled (or on-demand, or reactive). Each solution has their strengths and weaknesses. With configured mechanisms, traffic treatment (such as classification, priority, shaping, etc.) can be specified as well as the characteristics of the traffic to receive that treatment. An administrator would observe traffic patterns on the network, compare that with the desired state (based on business or operational needs), and then choose policies that allocate resources accordingly. Such mechanisms may work quite well for traffic such as HTTP, telnet, or FTP, which are tolerant to the variance in flow quality (jitter, packet reordering, etc.).

The outsourcing model is totally different. A policy enforcement device issues a request to ask for a decision for a specific request coming from a user, a program or a process. For example, the arrival of an RSVP message to a PEP requires a fast policy decision to avoid a long delay for an end-to-end set-up. The PEP may use COPS-RSVP to send a request to the PDP, soliciting for a policy decision.

Note that the outsourcing policy scheme differs with configuring policy scheme, but they are not mutually exclusive and operational systems may combine both.

The strength of signaling is that it enables the network to offer QoS guarantees, and to simultaneously be used efficiently. Without signaling, it is necessary either to compromise the quality of the guarantees, or to overprovision the network. In some networks, over provisioning may be a viable option. However, in other networks it may not. If the network manager wants to have the flexibility to not overprovision the network then, an end to end signaling must be available to be used for policy-based admission control decisions.

Signaling mechanisms can provide information beyond the QoS needs to handle the traffic. User information and application identification that could be hidden by IPsec can be provided, thus allowing higher quality information on the traffic.

One of the most difficult parts of PBN concerns policy translations: transformation of a policy from a representation or from a level of abstraction, to another representation or level of abstraction. For example, it may be necessary to convert a PIB data (Policy Information Base, e.g., a named data structure) to a command line format. In this conversion, the translation to the new representation is likely to require a change in the level of abstraction. Although these are logically distinct tasks, they are in most cases hidden in the acts of translating or converting or mapping. Therefore, policy conversion or policy mapping is an important problem that we do not look at in this paper.

3. EXAMPLES OF POLICY-BASED NETWORKING ARCHITECTURE

Policy-based systems may be used to manage and control different types of functionalities. In this section we will have a look at different examples where PBN may be applied. A first example concerns admission control schemes. These schemes are responsible for ensuring that the requested resources are available. Moreover, these schemes must take care of temporal constraints, identification and permission.

Policy-based admission control is able to express and enforce rules with temporal dependencies. For example, a group of users might be allowed to make reservations at certain levels only during off-peak hours. In addition, the policy-based admission control should also be able to support policies that take into account identity or credentials of users requesting a particular service or resource. For example, through a PBN scheme, an RSVP reservation request may be denied or accepted based on the credentials or identity supplied in the request.

A second example concerns Authentication, Authorization, Accounting (AAA) schemes. AAA deals with control, authentication, authorization and accounting of systems and environments. The schemes may be based on policies set by the administrator and users of the systems. The use of policy may be implicit or explicit. For example, a network access server can send dial-user credentials to an AAA server, and receives authentication that the user is who he claims, along with a set of attribute-value pairs authorizing various service features. Policy may be implied in both the authentication, which can be restricted by time of day, number of sessions, calling number, etc., and the attribute-values authorized.

A third example concerns quality of service (QoS). QoS refers to the ability to deliver network services according to the parameters specified in a Service Level Agreement. Quality of service is characterized by service availability, delay, jitter, throughput and packet loss ratio. At a network resource level, quality of service refers to a set of capabilities that allow a service provider to prioritize traffic, control bandwidth, and network latency. There are two different approaches to the quality of service on IP networks: Integrated Services (IntServ), and Differentiated Service DiffServ. IntServ approaches require policy control over the creation of signaled reservations, which provide specific quantitative end-to-end behavior for a flow. In contrast, DiffServ approaches require policy to define the correspondence between codepoints in the IP packet DS-field and individual per-hop behaviors to achieve a specified per-domain behavior. A maximum of 64 per-hop behaviors limit the number of classes of service traffic that can be marked at any point in a domain. These classes of service signal the treatment of the packets with respect to various QoS aspects such as flow priority and packet drop precedence. In addition, policy can be used to specify the forwarding of packets based on various classification criteria. The policy controls the set of configuration parameters and forwarding for each class in DiffServ, and the admission conditions for reservations in IntServ.

Another example is provided with VPN configuration. Let us first recall VPN meaning since the term has been widely used with a great deal of confusion. VPN is a set of terminal equipments that can communicate with each other. More formally, a VPN is defined by a set of administrative policies that control connectivity, quality of service, security, etc. among terminal equipment. A classical use of VPNs is security. Rather than impose the network manager to set security mechanisms for individual terminal equipment, policy-based management system can consolidate and synchronize access control lists and related policy information to promote a consistent security policy across the enterprise. For example, the network manager can use a policy-based management system to set policies for selection of tunneling protocols and to update client configuration instead of configuring each security device and each terminal equipment, making the management of the VPN system more scalable.

A last example of the use of PBN architecture on traffic engineering may be considered [10]. An IP traffic engineering policy could be applied for provisioning or allocating resources of an IP network. The allocation should be in correspondence with the quality of service negotiated by the terminal equipment. The use of COPS protocol to dynamically enforce traffic engineering policies yields the generic model shown in Figure 2.

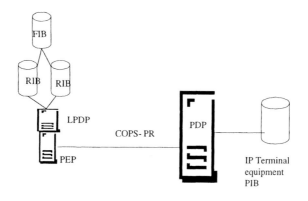

Figure 2 - A model of an IP traffic engineering policy enforcement scheme

As shown in Figure 2, the edge router is a policy enforcement point, which supports the IP Traffic Engineering (TE) client-type. The IP TE client-type is specified by the PEP to the PDP, and is unique for the area covered by the IP traffic engineering policy, so that the PEP can treat all the COPS client-types it supports as non-overlapping and independent namespaces. The router contains also a LPDP, which can store the routing processes that have been activated in the router. Within the context of enforcing an IP traffic engineering policy, the LPDP is expected to calculate and install the IP TE routes that comply with the QoS requirements expressed in the IP TE-related information that has been received by the PEP. The router has to contain instances of Routing Information Bases (RIB), according to the different routing processes that have been activated. No specific assumption was made about the actual number of RIB instances that can be supported by the router, since this is an implementation specific issue. The PEP has also to contain a Forwarding Information Base (FIB), which will store the routes that have been selected by the routing processes.

The enforcement of an IP traffic engineering policy is based upon the use of an IP TE policy server, the PDP that sends IP TE-related information to the PEP capability contained in the router. The IP TE-related information is stored and maintained in the IP TE Policy Information Base, which will be accessed by the PDP to retrieve and update the IP TE-related information whenever necessary.

The IP TE-related information is conveyed between the PDP and the PEP thanks to the establishment of a COPS-PR connection between these two entities. The COPS-PR protocol provides a named data structure (the PIB), to identify the type and purpose of the policy information that is sent by the PDP to the PEP for the provisioning of a given policy.

As in COPS-PR, IP traffic engineering policy information is described as a named data structure, a PIB. Here, the data structure is described as a collection of PRovisioning Classes. Furthermore, these classes contain attributes that actually describe the IP TE-related information that will be sent by the PDP to the PEPs. These attributes consist of the link and traffic engineering metrics that will be manipulated by the routing processes being activated in the routers to calculate the IP TE routes for a given destination, among other characteristics.

This approach clearly assumes that each service provider will have the ability to instantiate the contents of its own IP TE PIB, according to the routing policies that have been defined for forwarding the traffic within its domain, but also outside of its domain.

4. NEW EXTENSIONS TO POLICY-BASED ARCHITECTURE

While the focus of many early systems for policy-based networking has been the control of edge devices such as edge routers, firewalls, or gateways, future systems should have to account for end-user hosts as policy enforcement points. In fact, it is necessary to look at these terminal equipments as PEPs, both to provide finer-grained classification of traffic and to deal with traffic classification problems that can arise when traffic from the user terminal is encrypted. Problems with network congestion and QoS adaptation will be solved by enforcing policies at the terminal equipment, requiring this terminal to be well aware with regards to the network traffic it generates. We believe that in the future the enforcement points could not be edge routers except complicating the way to enforce the policy on these machines.

So, we think that COPS is able to take place in this new architecture as a homogenization element to take care directly of the terminal equipment in its quality of service, its mobility, its security, etc. We can define a new client type that would permit to interface directly the customer with the PDP.

To get a direct negotiation with the PDP, we have proposed an Internet draft [11] using COPS protocol for supporting SLS (Service Level Specification) negotiation. COPS-SLS is an extension of COPS protocol. The advantage when using COPS for SLS negotiation is the inherent flexible characteristic of COPS protocol. COPS may support multiple client-types. So, COPS-SLS protocol needs only to specify corresponding new objects used in this client-type (COPS-SLS). The client-handle object defined by the COPS protocol gives a mechanism for handling various requests in a single PEP. This capability will be used to handle several SLS negotiations from a single PEP.

The PEP in COPS-SLS is just a logical entity which requests network resources for itself or possibly on behalf of other entities. So, the client may be an end-host, or a gateway of a local network or another ISP. The model we have implemented is illustrated in Figure 3.

To negotiate a level of service, COPS-SLS has two phases: Configuration phase and Negotiation phase. The communication starts with the Configuration phase. The PDP uses the Configuration model to configure the Negotiation phase. After that, in the Negotiation phase, the client use the Outsourcing model to request a level of service with parameters conforming to the configuration installed in the Configuration phase. This organization in two phases makes the SLS negotiation dynamic. At any time, when the network sends a new configuration to the client, the Negotiation process will apply these policies in subsequent service level requests.

To negotiate a level of service, the client sends a request indicating its desired service level under the form of instances of PIB classes. Using PIB to represent SLS information makes COPS-SLS flexible and adapted to desired negotiation parameters of network providers. COPS-SLS protocol is designed to permit basic activities in SLS negotiation. The client can request, modify or terminate a level of service. The network can accept or reject a service level request, propose another service level to the client or degrade a service level when necessary.

With COPS-SLS protocol, it is easy for a company to install an Intranet with a policy-based control. This policy-based control may allow some applications to get a good quality of service and some others to be delayed. Packets of these applications may be given a very low level of priority within the company or even may be discarded as a non-appropriate traffic. This solution could be used as a basis for a security system using a firewall.

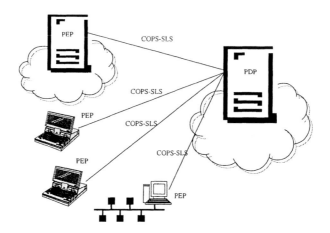

Figure 3 – The COPS-SLS model

COPS-SLS is suitable for the negotiation of SLS between network providers. A domain may negotiate with another domain to obtain a level of service for inter-domain communications. For example, a DiffServ domain may request another DiffServ domain to guarantee a level of service for all packets having a specific DSCP.

Another application of the policy-based architecture we developed concerns the management of the mobility of the terminal or the mobility of the user. COPS extension for Mobile IP policy registration control was proposed in [12]. This proposal deals with terminal mobility management. We introduce a policy-based architecture to support mobile user and mobile terminal registration, service portability, and QoS negotiation in fixed and wireless network access. The first challenge of this work is to define a policy based architecture to support user and terminal registration to achieve location management. The second challenge is to define a policy based architecture to support fixed or mobile service portability and QoS negotiation. To achieve these challenging goals, we introduce new components in the IETF policy-based architecture (e.g., Figure 4) and we introduce two COPS extensions called COPS-MU (Mobile User) and COPS-MT (Mobile Terminal), which define new policy objects to support user and terminal registration, service portability, and QoS negotiation.

QoS, Security, and Mobility Management 177

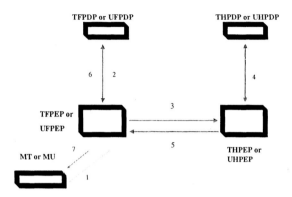

Figure 4 – The COPS-MU model

Figure 4 describes the new components used in the COPS-MU or COPS-MT architecture related to mobile terminal and mobile user. Some mobile IP terms used in this figure are explained below:
- ***TFPDP/UFPDP*** Terminal/User Foreign Policy Decision Point
- ***TFPEP/UFPEP*** Terminal/User Foreign Policy Enforcement Point
- ***THPDP/UHPDP*** Terminal/User Home Policy Decision Point
- ***THPDP/UHPDP*** Terminal/User Home Policy Enforcement Point.

The process represented by the sequence 1 to 7, in Figure 4, permits to treat registration, service portability, QoS assignment and mobility for terminal and user mobility [13].

Both techniques, COPS-SLS and COPS-MU, may be combined to avoid T/UFPEPs and T/UHPDPs devices in COPS-MU.

Associated with these policies, it is possible to address allocation schemes using DHCP protocol. Policies can dictate how sets of IP addresses are to be allocated and for what duration. It is also possible to address routing policies, VPN policies and many others IP protocols management schemes. Some extensions have been provided in [14].

5. CONCLUSION

This paper presented an introduction to policy-based management within an IP network. It also looked at different proposals to use efficiently policy-based management techniques. This management is related to translating high-level user needs to device-specific configuration. An important problem concerns the dynamic of the system. Adaptive policies could handle the changes in the network. Rules reflecting a change can be placed in the policy repository. The change may happen from a threshold or simply from the time of day. Agents on PEPs may indicate these changes to the PDP. The PDP decides about the set of policies that need to be applied. PDP may also choose the policy that satisfies the requirement of a high-level policy determined by a business needs. Another way to consider the problem is to support a policy discovery system that should determine the best policy to apply to a user requesting for a transmission. For example, the policy discovery system may decide to use IPsec if the flow has to traverse the Internet.

The policy is applied on an administrative domain and another challenge concerns interdomain policies. It would be interesting to see how the notion of policies applies across administrative domains. The use of COPS protocol is an available solution to correlate the policies to be chosen when a flow has to cross several administrative domains. An agent negotiation is another possibility to settle the policies to be applied.

Finally, policy monitoring has also to be determined to verify that the network is meeting the desired business needs. The monitoring system checks that the implementation of the policy complies with what was expected by the PDP. Indeed, the high-level policies reflect the SLAs, and the monitoring system must confirm that these SLAs are performed.

As a conclusion, we think that the most important function of policy-based networking systems is to simplify network management and operations in complex networks. These systems provide QoS, security, mobility and much more functions within a homogeneous way.

References

[1] RFC 2748 – D. Durham, J. Boyle, R. Cohen, S. Herzog, R. Rajan, A. Sastry, The COPS (Common Open Policy Service) Protocol, January 2000.

[2] RFC 2749 – J. Boyle, R. Cohen, D. Durham, S. Herzog, R. Rajan, and A. Sastry, COPS usage for RSVP, January 2000.

[3] RFC 2750 – S. Herzog, RSVP Extensions for Policy Control, January 2000.

[4] RFC 2751 – S. Herzog, Signaled Preemption Priority Policy Element, January 2000.

[5] RFC 2752 – S. Yadav, R. Yavatkar, R. Pabbati, P. Ford, T. Moore, S. Herzog, Identity Representation for RSVP, January 2000.

[6] RFC 2753 – R. Yavatkar, D. Pendarakis, R. Guerin, A Framework for Policy-based Admission Control, January 2000.

[7] RFC 2872 – Y. Bernet et R. Pabbati, Application and Sub Application Identity Policy Element for Use with RSVP, June 2000.

[8] RFC 2940 – A. Smith, D. Partain, J. Seligson, Definitions of Managed Objects for Common Open Policy Service (COPS) Protocol Clients, October 2000.

[9] RFC 3084 – K. Chan, J. Seligson, D. Durham, S. Gai, K. McCloghrie, S. Herzog, F. Reichmeyer, R. Yavatkar, and A. Smith, COPS Usage for Policy Provisioning, March 2001.

[10] C. Jacquenet, A COPS client-type for IP traffic engineering, Internet draft, http://www.ietf.org/internet-drafts/ draft-jacquenet-ip-te-cops-02.txt, June 2001.

[11] T.M.T. Nguyen, N. Boukhatem, Y. El Mghazli, N. Charton, G. Pujolle, COPS Usage for SLS negotiation (COPS-SLS), Internet Draft, <draft-nguyen-rap-cops-sls-02.txt>, April 2002.

[12] M. Jaseemuddin, A. Lakas, COPS usage for Mobile IP, Internet draft, draft-jaseem-rap-cops-mip-00.txt, October 2000.

[13] H. Chaouchi, G. Pujolle, COPS-MU: Policy based user mobility management, Proceeding IEEE Conference on Applications and Services In the Wireless Public Infrastructure, Evry, France, July 2001.

[14] T.M.T. Nguyen, N. Boukhatem, Y. Ghami Doudane, G. Pujolle, COPS SLS: A Service Level Negotiation Protocol for the Internet, IEEE Communications Magazines, May 2002.

A Multicast Routing Protocol with Multiple QoS Constraints

Li Layuan and Li Chunlin*
Department of Computer Science. Wuhan University of Technology,
Wuhan 430063.P. R. China
E-mail: jwtu @ public. wh. hb. cn

Abstract: The next generation Internet and high-performance networks are expected to support multicast routing with QoS constraints. To facilitate this, QoS multicast routing protocols are pivotal in enabling new member to join a multicast group. Multicast routing is establishing a tree which is rooted from the source node and contains all the multicast destinations. A multicast routing tree with multiple QoS constraints may be the tree in which the delay, delay jitter, packet loss and bandwidth should satisfy the pre-specified bounds. This paper discusses the multicast routing problem with multiple QoS constraints, which may deal with the delay, delay jitter, bandwidth and packet loss metrics, and describes a network model for researching the routing problem. It presents a multicast routing protocol with multiple QoS constraints (MRPMQ). The MRPMQ attempts to significantly reduce the overhead of constructing a multicast tree with multiple QoS constraints. In MPRMQ, a multicast group member can join or leave a multicast session dynamically, which should not disrupt the multicast tree. It also attempts to minimize overall cost of the tree, and satisfy the multiple QoS constraints and least cost's (or lower cost) requirements. In this paper, the proof of correctness and complexity analysis of the MRPMQ are also given. Simulation results show that MRPMQ is an available approach to multicast routing decision with multiple QoS constraints.

Keywords: Multicast routing ; protocol ; multiple QoS constraints ; QoS routing ; NP-complete.

---

*The work is supported by National Natural Science Foundation of China and NSF of Hubei Province.

1 INTRODUCTION

With the rapid development of Internet, mobile networks and high-performance networking technology, multicast routing has continued to be a very important research issue in the areas of networks and distributed systems. It attracts the interests of many people. Multicast services have been used by various continuous media applications. For example, the multicast backbone (Mbone) of the Internet has been used to transport real time audio/video for news, entertainment, video conferencing, and distance learning. The provision of Quality-of–Service (QoS) guarantees is of utmost importance for the development of the multicast services.

The traditional multicast routing protocols, e.g., CBT and PIM [1-4], were designed for best-effort data traffic. They construct multicast trees primarily based on connectivity. Such trees may be unsatisfactory when QoS is considered due to the lack of resources. Several QoS multicast routing algorithms have been proposed recently. Some algorithms [5-10] provide heuristic solutions to the NP-complete constrained Steiner tree problem, which is to find the delay-constrained least-cost multicast trees. These algorithms however are mot practical in the Internet environment because they have excessive computation overhead, require knowledge about the global network state, and do not handle dynamic group membership. Jia's distributed algorithm [5] does not compute any path or assume the unicast routing table can provide it . However, this algorithm requires excessive message processing overhead. The spanning join protocol by Carlberg and Crowcroft [1] handles dynamic membership and does not require any global network state. However, it has excessive communication and message processing overhead because it relies on flooding to find a feasible tree branch to connect a new member. QoSMIC [6], proposed by Faloutsos et al., alleviates but does not eliminate the flooding behavior. In addition, an extra control element, called Manager router, is introduced to handle the join requests of new members.

Multicast routing and its QoS-driven extension are indispensable components in a QoS-centric network architecture[15-17]. Its main objective is to construct a multicast tree that optimizes a certain objective function (e.g., making effective use of network resources) with respect to performance-related constraints (e.g., end-to-end delay bound, inter-receiver delay jitter bound, minimum bandwidth available, and maximum packet loss probability).

In this paper, we study the delay, delay jitter, bandwidth, and packet loss-constrained low cost QoS multicast routing problem which is known to be NP-complete, describe a network model for researching the routing problem, and present a multicast routing protocol with multiple QoS constraints (MRPMQ). In MRPMQ, a multicast group member can join or leave a multicast session dynamically. The MRPMQ can significantly reduce the overhead of constructing a multicast tree. The join of new member can have minimum overhead to other on-tree nodes or non-tree nodes, and only requires minimum adjustment to original tree and minimum resource reservation process. The protocol may search multiple feasible tree branches, and can select the best branch connecting the new member to the tree. It can also minimized the overall cost of the tree, and satisfy multiple QoS constraints and least cost's (or lower cost) requirements. The protocol can operate on top of some available unicast routing protocol.

The rest of the paper is organized as follows. Section 2 discusses the QoS multicast routing problem and descryibes a network model. Section 3 presents the MRPMQ. Section 4 gives the correctness proof and complexity analysis. Some simulation results are provided in Section 5. The paper concludes with Section 6.

2 NETWORK MODEL

As far as multicast routing is concerned, a network is usually represented as a weighted digraph $G = (V, E)$, where V denotes the set of nodes and E denotes the set of communication links connecting the nodes. $|V|$ and $|E|$ denote the number of nodes and links in the network, respectively, Without loss of generality, only digraphs are considered in which there exists at most one link between a pair of ordered nodes[13]. Associated with each link are parameters that describe the current status of the link.

Let $s \in V$ be source node of a multicast tree, and $M \subseteq \{V-\{s\}\}$ be a set of end nodes of the multicast tree. Let R be the positive weight and R^+ be the nonnegative weight. For any Link $e \in E$, we can define the some QoS metrics: delay function *delay* (e): $E \rightarrow R$, cost function *cost* (e): $E \rightarrow R$, bandwidth function *bandwidth* (e); $E \rightarrow R$, and delay jitter function *delay-jitter* (e): $E \rightarrow R^+$. Similarly, for any node $n \in V$, one can also define some metrics: delay function *delay* (n): $V \rightarrow R$, cost function *cost* (n): $V \rightarrow R$, delay jitter function *delay-jitter* (n): $V \rightarrow R^+$ and packet loss function *packet-loss* (n): $V \rightarrow R^+$. We also use $T(s, M)$ to denote a multicast tree, which has the following relations:

1) $delay\ (p\ (s,t)) = \sum_{e \in P(s,t)} delay\ (e) + \sum_{e \in P(s,t)} delay\ (n)$

2) $cost\ (T(s,M)) = \sum_{e \in T(s,M)} cost\ (e) + \sum_{e \in T(s,M)} cost\ (n)$

3) *bandwidth* $(p(s,t)) = \min\{bandwidth\ (e),\ e \in P(s,t)\}$.

4) *delay-jitter* $(p\ (s,t)) = \sum_{n \in P(s,t)} delay - jitter\ (e)$
   $+ \sum_{n \in P(s,t)} delay - jitter\ (n)$

5) *packet-loss* $(p\ (s,t)) = 1 - \prod_{n \in P(s,t)}(1 - packet\text{-}loss\ (n))$

where $p\ (s,t)$ denotes the path from source s to end node t of $T\ (s, M)$.

Definition 1. QoS-based multicast routing problem deals mainly with some elements: Network $G=(V,E)$, multicast source $s \in V$, the set of end nodes $M \subseteq \{V-\{s\}\}$, $delay(\cdot) \in R$, $delay\text{-}jitter(\cdot) \in R^+$, $cost\ (\cdot) \in R$, $bandwidth\ (\cdot) \in R$, and $packet\text{-}loss\ (\cdot) \in R^+$. This routing problem is to find the $T\ (s, M)$ which satisfies some QoS constraints:

1. Delay constraint: $delay\ (p\ (s,t)) \leq D_t$
2. Bandwidth constraint: $bandwidth\ (p\ (s,t)) \geq B$
3. Delay jitter constraint: $delay\text{-}jitter\ (p\ (s,t)) \leq J$
4. Packet loss constraint: $packet\text{-}loss\ (p\ (s,t)) \leq L$

Meanwhile, the *cost* $(T\ (s, M)$ should be minimum. Where D is delay constraint, B is bandwidth constraint, J is delay jitter constraint and L is packet loss constraint. In the above QoS constraints, the bandwidth is concave metric, the delay and delay jitter are additive metrics, and the packet loss is multiplicative metric. In these metrics, the multiplicative metric can be converted to the additive metric. For simplicity, we assume that all nodes have enough resource, i.e., they can satisfy the above QoS constraints. Therefore, we only consider the links' or edges' QoS constraints, because the links and the nodes have equifinality to the routing issue in question. Fig. 1 shows the network model. The characteristics of edge can be described by a fourtuple (D,J,B,C), where D,J,B and C denote delay, delay jitter, bandwidth and cost, respectively.

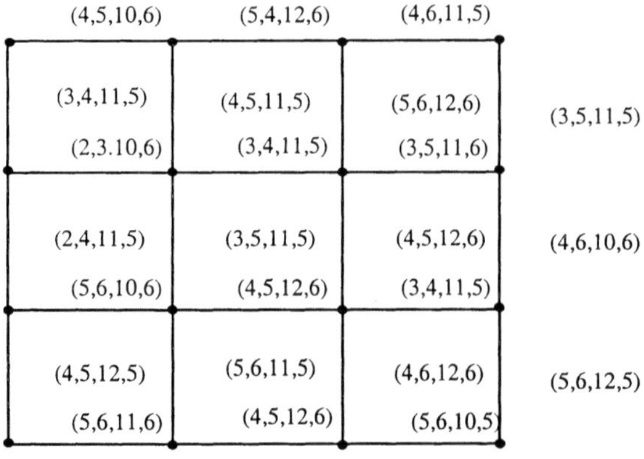

Fig. 1 An example of the network model

3   MRPMQ

3.1   Overview

MRPMQ can operate on top of some available QoS unicast routing protocol that pre-computes QoS paths. QOSPF could be used for the intra-domain routing operation. For inter-domain routing operation, it may require an available inter-domain unicast routing protocol. Though MRPMQ can works on top of OSPF like protocols, it is different from the MOSPF. In MRPMQ, a router (or a node) keeps a routing entry R (G, s, in, out), where $R.G$ is the group address, $R.s$ is the source address, $R.in$ is the incoming network interface, and $R.out$ is the set of outgoing network interfaces. Main control messages of MRPMQ can be described by the following Definition 2.

Definition 2. The set of control message can be defined as follows. JOINreq-a join request message sent towards source s by a new member that wishes to join a multicast group (G), JOINack-an accept acknowledgment sent downstream towards the new member by some node that accepts the join request, JOINnak-a rejection notification sent downstream towards the new member by some node that rejects the join request, and JOINpend-if the delay of a searching path does not satisfy the constraint D, and the next node (t_{j*}) is not the immediate upstream node of t_j, then t_j should add a pending information to the routing entry and marks t_{j*} as a possible upstream node.

A multicast group member may join or leave a multicast session dynamically. It is thus important to ensure that member join/leave will not disrupt the on-going multicast, and the multicast tree after member join/leave still satisfies the QoS requirements of all the on-tree receivers and remains near optimal. If a multicast tree is re-constructed from the source each time a member joins or leaves, on-tree nodes may not switch to the new tree simultaneously and the packets that are originally routed on the old multicast tree may be lost and have to be retransmitted. A seamless transition may thus not be achieved. Hence, in the case of member join/leave, a method that incrementally changes the $T(s,M)$ can be used. When a receiver leaves a multicast session, if the receiver is a leaf node of a multicast tree, it sends a leave message upstream. The leave message travels upstream along the on-tree branch until it reaches a fork node (i.e., a node with more than one downstream on-tree interface and/or with receivers on its directly attached subnet). Upon receipt of a leave message, an intermediate node releases the network resources. The rest of the multicast tree remains unchanged. On the other hand, if the leaving receiver is not a leaf node, it is not disconnected from the multicast tree. From this point onwards, this node simply relays incoming messages to the outgoing downstream interfaces. The ongoing session will not be affected.

When a receiver intends to join a multicast session, it sends a join request JOINreq to the source of the session with the information of its delay, delay jitter, bandwidth and cost. When a node receives a JOINreq from other node with IGMP, it will computes a path from s to itself, if the node is not the on-tree node. If a join request is accepted, a JOINack is sent downstream towards the new member. Otherwise if a join request is rejected, a JOINnak is also sent to the receiver. In MRPMQ, the source node forwards the QoS probing message to all receivers every ten seconds, so that QoS constraints information can be carried in the message, which can update current QoS requirements on time. In addition, MRPMQ also assumes each receiver knows all Link state information and the address of the multicast group in its intra-domain.

3.2 Detailed description

This section will describe some details of MRPMQ, the emphasis is laid on the join process, multicast tree construction and multiple constraints.

In MRPMQ, a searching tree is formed incrementally. When a receiver, such as the leaf node t_i (see Fig.2), wants to join the multicast group, it send a JOINreq to intermediate node (router) t_1. When t_1 receives the JOINreq message from t_i, and it is already the on-tree node, it will make an eligibility test. It check whether or not the QoS requirement of the new member as well as the existing QoS guarantees to the other on-tree members, can be fulfilled. Suppose the path delay and delay jitter from t_1 to t_0 are $d(1,0)$ and $d_j(1,0)$, respectively. Similarly, the path delay and delay jitter from t_i to t_1 should also be $d(i,1)$ and $d_j(i,1)$, respectively. Let $bw(t_1,t_0)$ be the bandwidth of lind from t_1 to t_0, and $bw(t_i,t_1)$ be the bandwidth of link from t_i to t_1. Recall that the delay, delay jitter and bandwidth constraints are D, J and B, respectively. The t_1 will check if

$$(d(i,1)+d(1,0) \leq D) \wedge (dj(i,1)+dj(1,0) \leq J) \wedge (bw(t_i,t_1) \wedge bw(t_1,t_0) \geq B)$$

it will add the corresponding *out* to the routing entry, and transfers a JOINack message to t_i. Otherwise, t_1 will transfer JOINnak message to t_i. In Fig.2, other nodes are the on-tree nodes of the multicast group.

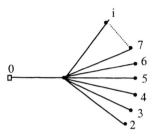

Fig. 2 t_i requires to join multicast group

If $(d(i,1)+d(1,0)>D) \wedge (dj(i,1)+dj(1,0) \leq J)$ then there are two possible cases.

Case 1 If next hop in the JOINreq is just the immediate upstream node $t_{1'}$ (see Fig.3), t_1 can transfer the JOINreq from the new member t_i to $t_{1'}$, and add *out* to the corresponding routing entry and mark it as pending. In the case, $t_{1'}$ may accept the join request or transfer the join request upstream.

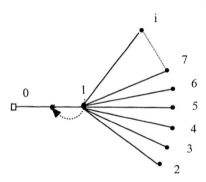

Fig. 3 $t_{1'}$ is the immediate upstream node

Case 2 If next hop is not the immediate upstream node, such as t_{1*} shown in Fig.4, and if

$$(d(0,1')+d(1',1) \geq D-d(1,i)) \wedge (dj(0,1')+dj(1',1) \geq J-dj(1,i))$$

then t_1 will transfer the JOINreq to t_{1*}, and add a pending routing entry and mark t_{1*} as upstream rode. If t_{1*} transfer JOINack message to t_1, t_1 will forward a pruning message to $t_{1'}$. This case will produce a switch from the original path $(t_0 \rightarrow t_{1'} \rightarrow t_1)$ to the new path $(t_0 \rightarrow t_{1*} \rightarrow t_1)$.

Under this situation, the t_{1*} should send a QoS probing message with the new QoS requirements to all downstream nodes.

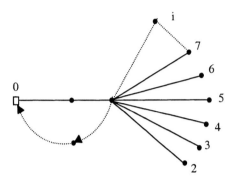

Fig. 4 Switch to new path ($t_0 \rightarrow t_{1*} \rightarrow t_1$)

If $d(0,1')+d(1',1) \leq D$ or $dj(0,1')+dj(1',1) \leq J$, t_1 will compute a new path. The new path should satisfy delay constraint $\min[d(0,1), D-d(1,i)]$ and delay jitter constraint $\min[dj(0,1), J-dj(1,i)]$. If so, t_1 will receive the JOINack message, otherwise it will receive the JOINnak message. The situation can be shown in Fig.5.

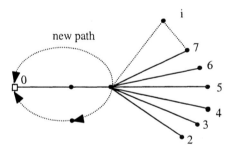

Fig. 5 Compute the new path

The Join procedure of MRPMQ can be formally described as follows.

1. if a new member (ti) wishes to join a $T(s,M)$ the new member sends JOINreq to some neighbor t_j
2. if $(d(s,*)+d(j,i) \leq D) \square (d_j(s,*)+dj(j,i) \leq J) \square (bw(t_u,t_v) \geq B)$; {where $d(s,*)$ and $d_j(s,*)$ are the delay sum and the delay jitter sum from the source s to all downstream nodes of a path, respectively, but except for the last a pair of nodes. The u and v are the sequence numbers between two adjacent nodes on path from source to the new member}
   $\rightarrow t_j$ transfers JOINack to t_i
   fi

```
       if bw (t_u,t_v)<B→
          remove e (u,v) from G
       fi
```
3. if $(d(s,*)+d(j,i)>D) \Box (dj(s,*)+dj(j,i)>J)$→
   ```
   if the next hop is the immediate upstream node (t_j' of t_j)
      t_j transfers JOINreq to t_j'→
      t_j adds JOINpend for t_i to the forwarding entry
      t_j' transfers JOINack(or JOINnak) to t_j
   fi
   if the next hop is not the immediate upstream node.
      if (d(s,*) ≥ D-d(j,i)) ∧ (dj(s,*) ≥ J-dj(j,i))→
         t_j transfers JOIN reg to t_j*
         t_j adds the routing entry
         marks t_j* as upstream node
         t_j* transfers JOINack(or JOINnak)to t_j
         if t_j receives JOINack→
            t_j forwards a pruning msg to t_j'
         fi
      fi
   fi
   ```
4. if $(d(s,*)+d(j,i)\le D) \Box (dj(s,*)+dj(j,i)\le J)$→
   ```
   t_j computes a new path
   if (d (p (s,i))=min[d (s,*), D-d (j,i)]) ∧
      (dj(p (s,j))=min[dj (s,*), J-dj (j,i)])→
      t_j receives JOINack
   fi
   t_j receives JOINnak
   fi
   ```

We can use the following example to show how the MRPMQ works and how the multicast tree is constructed in a distributed fashion. Fig, 6 is a network graph. In this example, node t_0 is the multicast source. The t_4, t_9, t_{14}, t_{19} and t_{24} are the joining nodes. Recall that characteristics of network's edge can be described by a fourtuple (D,J,B,C). In this example shown in Fig 6, suppose delay constraint $D=20$, delay jitter $J=30$ and bandwidth constraint $B=40$, The t_4 wishes to join the group, it computes the paths according to the multiple QoS constraints: the path $(t_0 \to t_1 \to t_2 \to t_3 \to t_4)$. Path $(t_0 \to t_1 \to t_2 \to t_3 \to t_8 \to t_4)$ and path $(t_0 \to t_1 \to t_2 \to t_7 \to t_8 \to t_4)$ do not satisfy the delay constraint. The path $(t_0 \to t_6 \to t_7 \to t_8 \to t_4)$, path $(t_0 \to t_5 \to t_6 \to t_7 \to t_8 \to t_4)$ and path $(t_0 \to t_1 \to t_6 \to t_7 \to t_8 \to t_4)$ satisfy the delay constraint, the delay jitter constraint and the bandwidth constraint. Furthermore, the path $(t_0 \to t_1 \to t_6 \to t_7 \to t_8 \to t_4)$ has minimum cost among these paths. Therefore, the join path should be the path $(t_0 \to t_1 \to t_6 \to t_7 \to t_8 \to t_4)$. The bold lines of Fig.7(a) show the tree when t_4 has joined the group. When t_9 joins the group, it computes a path $(t_0 \to t_1 \to t_6 \to t_7 \to t_8 \to t_9)$ which should satisfy delay, delay jitter and bandwidth constraints, and also have minimum cost. The JOINreq is accepted at t_8. The bold lines of Fig.7(b) show the tree when t_9 has joined the group. When t_{14} join the group, it computes the paths with multiple QoS constraints. The path $(t_0 \to t_1 \to t_6 \to t_7 \to t_{12} \to t_{13} \to t_{14})$ does not satisfy the delay jitter constraint. The path $(t_0 \to t_5 \to t_6 \to t_7 \to t_8 \to t_{13} \to t_{14})$ does not satisfy delay and delay jitter constraints. The path $(t_0 \to t_6 \to t_7 \to t_{12} \to t_{13} \to t_{14})$ and path $(t_0 \to t_5 \to t_6 \to t_7 \to t_{12} \to t_{13} \to t_{14})$ satisfy delay, delay jitter and bandwidth constraints. The later has the lower cost. Therefore, the join path should be the path $(t_0 \to t_5 \to t_6 \to t_7 \to t_{12} \to t_{13} \to t_{14})$. Meanwhile, t_6 should prune off from original parent t_1, the resulted tree is shown in Fig.7(c) (see the bold lines of Fig.7(c)). The tree after t_{19} joins the group is also shown in Fig.7(d). When t_{24} joins the group, it computes the join paths. If t_{18} receives JOINreg from t_{24}, it find out that the existing path $(t_0 \to t_5 \to t_6 \to t_7 \to t_{12} \to t_{13} \to t_{18})$ does not satisfy the delay constraint for new member t_{24}, whilethe new path $(t_0 \to t_5 \to t_6 \to t_7 \to t_{12} \to t_{17} \to t_{18})$ does not satisfy the delay jitter constraint for t_{24}. The t_{18} computes a new feasible path with delay constraint, which is given by

$$d(p(s,j)) = \min[d(s,*), D-d(j,i)]$$
$$= \min[(d(0,5)+d(5,6)+d(6,7)+d(7,12)+d(12,13)+d(13,18)), D-d(18,24)]$$
$$= \min[19,18]$$
$$= 18$$

and delay jitter constraint, which can be given by

$$dj(p(s,j)) = \min[dj(s,*), J-dj(j,i)]$$
$$= \min[(dj(0,5)+dj(5,6)+dj(6,7)+dj(7,12)+dj(12,13)+dj(13,18)), J-dj(18,24)]$$
$$= \min[28,28]$$
$$= 28$$

Thus, this new feasible path should be path$(t_0 \to t_6 \to t_7 \to t_{12} \to t_{13} \to t_{18})$. The t_6 should prune off from old parent t_5, and the final tree can be shown in Fig.7(e)(see the bold lines of Fig.7(e)).

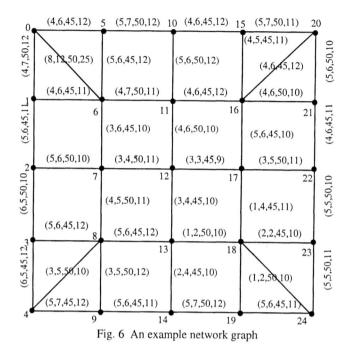

Fig. 6 An example network graph

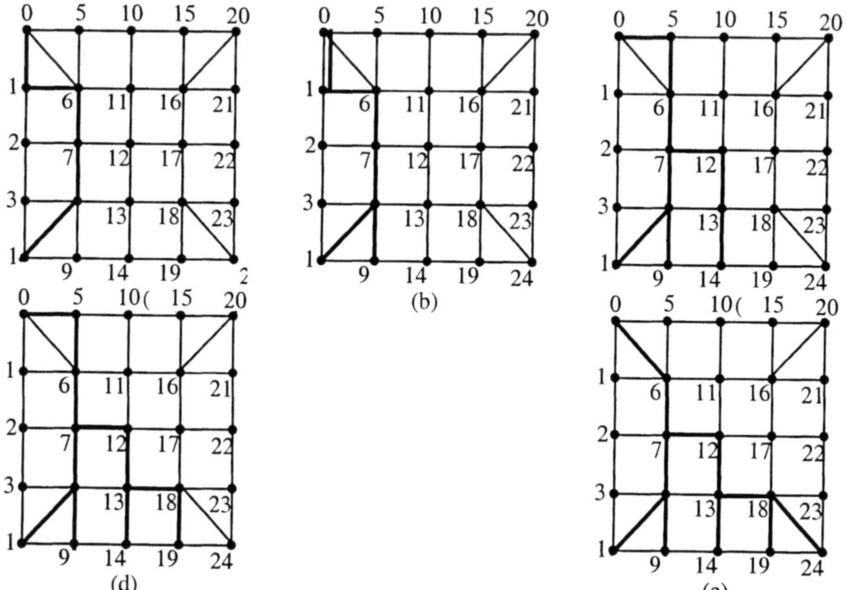

Fig. 7 Constructing multicast tree

The loop-free routing for the above protocol can be achieved by maintaining a searching tree at any time.

4 THE CORRECTNESS AND COMPLEXITY ANALYSIS

4.1 The correctness proof

Theorem 1. If a path from a new member to $T(s,M)$ has sufficient resources to satisfy the QoS constraints and has minimum cost, it searches only one path.

Proof. Note that a necessary condition for multiple paths to be searched is a single path does not satisfy the QoS constraints, such as $(d(p(s,j)) \neq \min[d(s,*), D-d(j,i)]) \wedge (dj(p(s,j)) \neq \min[dj(s,*), J-dj(j,i)])$. However, if sufficient resources are available on every link and node of the path, no node forwarding JOINreg will ever enter the multiple paths search state. Thus, the above theorem holds.

Lemma 1. Whenever during the routing process, all paths being searched form a $T(s,M)$ structure.

Proof. The paths being searched will be marked by the routing entries at the nodes. In MRPMQ, any routing entry has a single out interface and one or multiple in interfaces. Hence, the nodes will form a searching tree structure. This tree is just a $T(s,M)$.

Theorem 2. An available and feasible path found by MRPMQ is loop-free.

Proof. This Theorem follows directly from the above Lemma 1.

Lemma 2. If MRPMQ terminates without an available and feasible path, all nodes out of $T(s,M)$ are ether initial state or in failure state.

Proof. MRPMQ terminates without success only when the new member's JOINreg is rejected, i.e., it changes into the failure state. Since the new member is leaf node of the searching tree, when it changes into the failure state, all nodes in the searching tree must be in the failure state. The nodes outside the searching tree remain in the initial state.

Theorem 3. MRPMQ can find an available and feasible path if one exists.

Proof. This theorem can be proved by contradiction. Suppose MRPMQ fails while an available and feasible path does exist. Let $e(i,j)$ be the first link in the path that the protocol did not explore. Since $e(i,j)$ is the first unexplored link of the path, t_i must have received a request message from the previous link or t_i is the new member issuing the request message. In either case, t_i is not in the initial state. Therefore, t_i is in the failure state by Lemma 2. which requires t_i to explore all outgoing links including $e(i,j)$. It contradicts the assumption that $e(i,j)$ is not explored.

4.2 Complexity Analysis

The complexity of the QoS-based multicast routing protocol can be analyzed in terms of the computation complexity and the number of messages needed to construct a multicast tree. The former mainly concerns the computation overhead needed to find paths and construct the multicast tree. The later mainly involves the overhead of message exchange needed to build the multicast tree. In MRPMQ, route computation can generally be made by the end node. If the join path is computed on-demand, the complexity depends on the unicast protocol. If QoS metrics are delay and bandwidth, there exist QoS routing heuristics which are $O(|V| \times |E|)$, where $|V|$ is the number of nodes and $|E|$ is the number of edges in a network. For most networks, $|E|=O|V|$, hence the complexity is $O(|V|^2)$. For a multicast group with $|M|$ members, the computation overhead is $O(|V|^2|M|)$. The study shows that computation complexities of CSPT and BSMA [8] are $O(|E| \log |V|)$ and $O(|V|^3 \log|V|)$, respectively. The computation complexity of MRPMQ is $O(|V|^2 |M|)$. In overhead of message exchange, MRPMQ requires two messages, i.e., JOINreq and JOINack or JOINnak. This means that a multicast group with $|M|$ members deals with $2|M|$ message. A JOINreg will be processed by all K hops along the way up to the node where it is accepted or rejected, hence the overhead of message processing for joining $|M|$ members is $K.2|M|$. Jia's algorithm [5] also requires two control messages, however, each message actually includes full information of a multicast tree, and thus its message processing is more complex than in MRPMQ. QoSMIC[6] requites a considerable number of search of BID-ORDER messages to join a new member, all messages need processing at both the new member as well as other nodes. The study shows that the average message processing overheads to construct the multicast tree of MRPMQ, Jia's algorithm, and QoSMIC (centralized or distributed) are $K.2|M|$, $K.2|M|$, $|M| (w \cdot (w-1)^{(y-1)}+c-k) \cdot x$ (centralized QoSMIC) and $|M| (w \cdot (w-1)^{(y-1)}+|T|) \cdot x$ (distributed QoSMIC), respectively, where the x factor is added to reflect the fact that messages have to be processed at more than one node, w is the average degree of a node, y is maximum TTL used for search, $|T|$ is tree size, c is number of candidates for BID-ORDER session and x depends on the topology and y, while $2 \leq x \leq 1+K$.

5 SIMULATIONS

The effectiveness and availability of MRPMQ are studied by simulations. In the simulations experiment, four algorithms are simulated : MRPMQ, CSPT, BSMA and KMB [20]. KMB applies Prim's minimum spanning tree algorithm to the complete distance graph. This heuristic find a tree whose cost is within twice the cost of the corresponding Steiner tree.

Network topologies used in the simulations are deliberately manipulated to simulate wide area sparse networks. A large network is likely to be loosely interconnected [11,14]. An *n*-node graph is considered to be sparse when less than 5% of the possible edges are present in the graph. The network graphs used in the simulations are constructed by the Waxman's random graph model[12].

In this random graph, the edge's probability can be

$$P_e(u,v) = \beta \exp\left(-\frac{d(u,v)}{\alpha L}\right)$$

where $d(u,v)$ is geometric distance from node u to node v, L is maximum distance between two nodes, parameter α can be used to control short edge and long edge of the random graph, and parameter β can be used to control the value of average degree of the random graph.

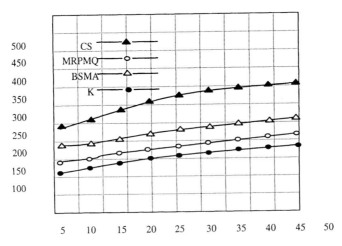

Fig. 8 Network cost vs. group size

In the simulations, we compare the quality of routing trees by their network cost[18]. The network cost is measured by the mean value of the total number of simulation runs. At each simulation point, the simulation runs 80 times. Each time the nodes in the group G are randomly picked out from the network graph. The network cost is simulated against two parameters: delay bound D and group size. In order to simulate the real situations, group size are always made less than 20% of the total nodes, because multicast applications running in a wide area network usually involve only a small number of nodes in the network, such as video conference systems, distance learning, co-operative editing systems, etc[19,21,22].

Fig. 8 shows the network cost versus group size. In this round of simulations, the network size is set to 300 and D is $d_{max}+3/8d_{max}$. From Fig. 8, we can see wthen group size grows, the network cost produced by MRPMQ, BSMA and KMB increases at a rate much lower than CSPT. The MRPMQ performs between BSMA and KMB. BSMA, KMB and the proposed MRPMQ can produce trees of comparable costs.

Fig. 9 is the network cost versus D. During this round of simulations, the network size is fixed at 300 nodes, group size is 20. We define the smallest meaningful value of D as $d_{max}=\max(\{d_u|\text{for any } u \in G: d_u \text{ is the delay on the shortest path from } s \text{ to } u\})$. D starts from d_{max}. With an even smaller D, there does not exist such a routing tree which satisfies the delay bound D. D is incremented by $d_{max}/8$ each time. The increment of $d_{max}/8$ is selected to maximally capture the trend of network cost against the change of D after many simulation runs. Since for each simulation run, G is different, thus d_{max} is different. The D values on the X-axis are the mean values of D in all runs.

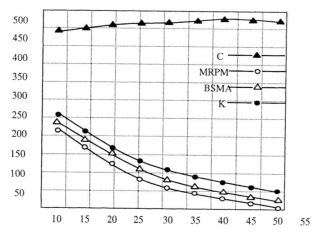

Fig. 9 Network cost vs. delay bound

From Fig.9, it can be seen that the network cost of the CSPT algorithm is on the top and almost does not change as D increases. This is because the generation of the shortest path tree does not depend on D. Of the remaining three algorithms, the proposed MRPMQ has the lowest cost. From Fig.9, we can also see that tree costs decrease for MRPMQ, BSMA and KMB algorithms as the delay bound is relaxed. This shows all three schemes indeed can reduce the cost when delay bound is relaxed. From Fig.8 and Fig.9, one can see that MRPMQ, BSMA and KMB algorithms can produce trees of comparable costs. However, compared with BSMA and KMB algorithms, the proposed MRPMQ has the advantage of being fully distributed and allowing incremental tree build-up to accommodate dynamic joining new members. Furthermore, the MRPMQ is much less costly in terms of computation cost and in terms of cooperation needed from other network nodes compared with other schemes.

6 CONCLUSION

In this paper, we discuss the multicast routing problem with multiple QoS constraints, which may deal with the delay, delay jitter, bandwidth and packet loss metrics, and describe a network model for researching the routing problem. We have presented a multicast routing protocol with multiple QoS constraints (MRPMQ). The MRPMQ can significantly reduce the overhead of establishing a multicast tree. In MRPMQ, a multicast group member can join or leave a multicast session dynamically, which should not disrupt the multicast tree. The MRPMQ also attempts to minimize overall cost of the tree. This protocol may search multiple feasible tree branches in distributed fashion, and can select the best branch connecting the new member to the tree. The join of new member can have minimum overhead to on-tree or non-tree nodes. The correctness proof and complexity analysis have been made. Some simulation results are also given. The study shows that MRPMQ is an available and feasible approach to multicast routing with multiple QoS constraints. Further work will investigate the protocol's suitability for inter-domain multicast and hierarchical network environment.

7 ACKNOWLEDGMENT

The work is supported by National Natural Science Foundation of China and NSF of Hubei Province.

8 REFERENCES

[1] K. Carberg and J. Crowcroft. "Building shared trees using a one-to-many joining mechanism". *ACM Computer Communication Review.* Jan. 1997, pp. 5-11.

[2] T. Ballardie, P. Francis and J. Crowcroft, "An architecture for scalable inter-domain multicast routing," *ACM SIGCOMM,* Sept. 1993, pp. 85-95.

[3] Li Layuan and Li ChunLin, "The QoS routing algorithm for ATM networks," *Computer Communications*, NO.3-4, Vol.24, 2001, pp. 416-421.
[4] Li Layuan. "A formal specification technique for communication protocol". *Proc of IEEE INFOCOM*, April. 1989, pp. 7481.
[5] X. Jia. "A distributed algorithm of delay-bounded multicast routing for multimedia applications in wide area networks," *IEEE/ACM Transactions on Networking*, No.6, Vol.6, Dec. 1998, pp. 828-837.
[6] M. Faloutsos, A. Banerjea, and R. Pankaj, "QoSMIC: Quality of Service sensitive multicast internet protocol," *SIGCOMM'98*. September 1998.
[7] L. Zhang, S. Deering, D. Estrin, S. Shenker, and D. Zappala, "RSVP: A new Resource ReSerVation Protocol," *IEEE Network*. Sept. 1993.
[8] Q. Zhu, M. Parsa, and J. J. Garcia-Luna-Aceves, "A source-based algorithm for delay-constrained minimum-cost multicasting," *Proc. IEEE INFOCOM 95*, Boston, MA, April 1995.
[9] Y. Xiong and L.G. Mason, "Restoration strategies and spare capacity requirements in self-healing ATM networks," *IEEE Trans on Networks*, Vol. 7, No. 1, Feb, 1999, pp. 98-110.
[10] S. Chen and K. Nahrstedt, "Distributed QoS routing in ad-hoc networks," *IEEE JSAC*, special issue on ad-hoc networks, Aug. 1999.
[11] Li Laywan. "The routing protocol for dynamic and large computer networks," Journal of computers, No.2, Vol.11, 1998, PP. 137-144.
[12] B. M. Waxman. "Routing of multipoint connections," *IEEE Journal of Selected Area in Communications*, Dec. 1998, pp. 1617-1622.
[13] R. G. Busacker and T. L. Saaty, Finite Graphs and Networks: An introduction with applications, McGraw-Hill, 1965.
[14] Li Layuan and Li Chunlin, "A routing protocol for dynamic and large computer networks with clustering topology," *Computer Communication*, No.2, Vol. 23, 2000, pp. 171-176.
[15] D. G. Thaler and C. V. Ravishankar, "Distributed center-location algorithms," *IEEE JSAC*, Vol. 15, April 1997, pp. 291-303.
[16] I. Cidon, R. Rom, and Y. Shavitt, "Multi-path routing combined with resource reservation," IEEE INFOCOM'97. April 1997, pp. 92-100.
[17] J. Mog. "Multicast routing exeensions to OSPF," RFC 1584. march. 1994.
[18] Li Layuan and Li Chunlin, "The QoS-based routing algorithms for high-speed networks." *Proc of WCC*, Aug. 2000, pp-1623-1628.
[19] Roch A. Guerin and Ariel Orda, "QoS routing in networks with inaccurate information: Theory and algorithms," *IEEE/ACM Trans. On Networking*, No.3, Vol.7. June. 1999. pp. 350-363.
[20] Bin Wang and Jennifer C. Hou, "Multicast routing and its QoS extension: Problems, algorithms, and protocols," *IEEE Network*, Jan/Feb, 2000. pp. 22-36.
[21] Li layuan and Li Chunlin, Computer Networking, National Defence Industry Press, Beijing, 2001.
[22] Moses Charikar, Joseph Naor and Baruch Schieber, Resoutce optimization in QoS multicast routing of real-time multimedia, *Proc of JEEE INFOCOM*. 2000. pp. 1518-1527.

Anonymous Internet Communication Based on IPSec

Ronggong Song and Larry Korba
Institute for Information Technology
National Research Council of Canada
E-mail: {Ronggong.Song, Larry.Korba}@nrc.ca

Abstract: Network approaches for anonymous communication have been extant for some time. Unfortunately, there are limitations with these approaches. In this paper, we first expose the limitations of existing anonymous communication networks. We then present an anonymous Internet communication technique based on IPSec. Our technique provides bi-directional, real-time anonymous Internet communication that is resistant traffic analysis for any TCP/IP applications. We describe the signaling protocols and the implementation of our architecture. Our technique is easily implemented in security gateways or IP router that supports IPSec.

Keywords: Internet security, privacy, anonymous Internet, traffic analysis.

1 INTRODUCTION

Privacy is becoming a critical issue on the Internet. Users feel that one of the most important barriers to using the Internet is the fear of having their privacy violated. Governments around the world have introduced legislation placing requirements upon the way in which personal information is handled. In attempt to provide some technical solutions within the privacy void, several network-based privacy-enhancing technologies have been developed in recent years. Some examples of these technologies include: MIX-Network [1,4], Onion Routing [7,11], Crowds System [12], Freedom Network [3], etc.

It is a difficult to achieve anonymity for real-time services in the Internet (e.g. Web Access). Some limitations of the above networks are described in

[2]. In addition, another important limitation is they are not implemented at the IP layer in such a way that some important security protocols cannot be available, for example, they all cannot support IPSec [8], and furthermore, the Crowds System cannot support SSL [6] at the endpoints. However, IPSec and SSL have become important security protocols, especially for e-commerce applications. Thus, an anonymous Internet at the IP layer, which can support any TCP/IP applications, becomes desiderata.

Our anonymous Internet proposal is based on the IPSec tunneling technique. There are two main reasons. First, the IPSec technique has become a popular security technique, especially for VPN application, and many security gateways, firewalls and routers support IPSec technique. Another reason is that it enables our proposal to support all TCP and UDP applications.

Our principles are as follows. First, our anonymous Internet comprises a set of anonymous Internet nodes (AINs) that support IPSec, such as security gateways, firewalls and routers, etc. The connection between two neighboring AINs is supported by the IPSec tunneling technique and built previously. We call the connection a permanent virtual tunnel (PVT). A creating signaling protocol then copies the PVT as a temporary virtual tunnel (TVT) with a different security parameter index (SPI), and creates an anonymous IP-datagram virtual tunnel (AIVT) from the entrance node to the exit node by connecting these TVTs. The signaling is transmitted through the PVTs. A nested symmetrical encryption channel is established by employing public key system during the AIVT. The end-to-end data then is encrypted using the nested symmetrical encryption algorithm and sent through the AIVT. Final, a destroying signaling protocol is used to destroy the AIVT and TVTs after the session.

The other advantages for our technique are as follows. First, unlike other anonymous communication networks, it is independent of all applications over IP. Unlike Onion Routing and MIX-Network, packet loss causes a problem of network backlog and cascading retransmits, since the nodes talk to each other via TCP in these networks. The end-to-end TCP is not available in them. On the other hand, unlike Freedom Network, in our anonymous Internet architecture, the source address and destination address of the user's data are changed through every node, but every node only needs one public IP address because we use the different IP tunnels to distinguish the different end-to-end IP-datagram streams. Thus, it supports both IPv4 and IPv6.

The rest of the paper is organized as follows. Background describing IPSec and anonymous communication networks are briefly reviewed in the next section. In Section 3, some notations are defined. In Section 4, anonymous IP routing protocols are proposed based on IPSec, including the creating PVT and AIVT protocols, and destroying AIVT protocol. In Section 5, the implementation of the anonymous Internet is concisely described. In Section 6, the vulnerabilities in our architecture are discussed. In Section 7, some concluding remarks and directions are presented for further research.

2 BACKGROUND

2.1 IPSec Architecture

The IPSec architecture is the most advanced effort in the standardization of Internet security. IPSec can be used to protect an IP layer path between a pair of end-systems or hosts, between a pair of intermediate systems - called security gateways.

IPSec consists of the following components:

- Two security protocols: the IP Authentication Header (IP AH) [9] and the IP Encapsulating Security Payload (IP ESP) [10] that provide the basic security mechanisms within IP;
- Security Associations (SA) that present the set of security services and parameters negotiated on each security IP path;
- Algorithms for authentication, encryption and integrity.

IP AH and IP ESP may be applied alone or in combination with each other. Each protocol can operate in one of two modes: transport mode or tunnel mode. In transport mode, the security mechanisms are applied only to the upper layer data, and the information contained in the IP header is left unprotected. In tunnel mode, both the upper layer data and the IP header are protected.

IP AH provides data origin authentication, data integrity and replay detection for IP datagrams. Except for these functions, IP ESP also provides data confidentiality services. But unlike the AH authentication data field, the authentication covers only the ESP header, the ESP payload and the padding fields of the IP datagram in the ESP data field. The IP header is never protected by the ESP authentication service. Thus, in the cases where the data integrity and data confidentiality of the entire IP datagram are required, it would be better to use IP ESP+AH.

A Security Association represents an agreement between two IP nodes on a set of security services to be applied to the IP traffic stream between these nodes. Each SA is associated with AH or ESP services but not both. IPSec provides end-to-end security and VPN services.

2.2 Anonymous communication networks

The primary goal of the anonymous communication network is to protect user anonymous communication against traffic analysis. Chaum first proposes an anonymous communication network: MIX-Network, for supporting anonymous e-mail services.

Based upon Chaum's MIX-Network, Dai [5] has described a theoretical architecture that would provide private protection against traffic analysis based on a distributed system of anonymous packet forwarders. He calls it Pipenet. Pipenet consists of a cloud of packet forwarding nodes distributed around the Internet, and packets from a client would be multiply encrypted and flow through a chain of these nodes. Pipenet is an idealized architecture and has never been built. Pipenet's mortal disadvantage is that its packet loss or delay is extremely large.

Like Pipenet architecture, Onion Routing provides a more mature implementation against traffic analysis. It also provides bi-directional real-time communications. Its disadvantages include that packet loss causes a problem, and does not support the security services such as IPSec, VPN, etc.

Freedom Network is another similar technique. It works in a very similar way as compared to the Onion Routing.

The Crowds System is based on a very different principle. A user sends a message with a certain probability into the Internet. Otherwise he forwards the messages to another randomly selected user. That user does the same things and so on. Unfortunately, the Crowds System doesn't support the security services such as IPSec, SSL, etc.

The above networks are not implemented at the IP layer in such a way that their applications have some limitations. Proposals for anonymous Internet at IP layer are in process now. Our proposal is to use the maturing IPSec techniques for supporting anonymous communication at IP layer.

3 TERMINOLOGY

Notations used in the paper are defined as follows.

- *AIN:* Anonymous Internet Node. It is a node such as security gateway, firewall or router supporting IPSec tunneling technique, converting the IP header of the IP datagram into another IP header for hiding the original IP header, and forwarding the data stream from one IP tunnel to another IP tunnel.
- *PVT:* Permanent Virtual Tunnel. This is an IP ESP+AH tunnel between two AINs that is created previously by Internet Key Exchange protocol

(IKE) or manual. It is a long-term tunnel depending on its security policy, and supports signaling services for our anonymous Internet.
- *TVT:* Temporary Virtual Tunnel. It is an IP AH tunnel between two AINs, and has the same parameters as the IP AH of the PVT except for SPI. It is a part of an AIVT, and supports data forwarding services for the user's datagrams.
- *AIVT:* Anonymous IP-datagram Virtual Tunnel. It comprises several TVTs that form the AIVT from the entrance AIN to the exit AIN. Note that an AIVT may include several UDP and TCP sessions with the same end-to-end IP addresses. A new AIVT is created for a new end-to-end IP-datagram stream by our creating AIVT protocol. An old AIVT is destroyed after a destroying signal is sent or the AIVT is expired.
- *SPI:* Security Parameter Index. It is a random value used in combination with the destination IP address to identify the security association for that datagram.
- $E_{PK_i}(K_i)$: The symmetrical key is encrypted with the AIN_i's public key PK_i, e.g. RSA.
- $E_{K_i}(M)$: The message M is encrypted with the symmetrical key K_i, e.g. DES.

4 ANONYMOUS IP ROUTING PROTOCOLS

4.1 Anonymous Internet Topology

An anonymous Internet consists of some AINs. The connection between two AINs is a PVT created by the IKE protocol or manual previously. Every AIN only accepts the data stream from its customers and some AINs that have a PVT or TVT with this AIN. The AIN then forwards the data stream to the next AIN according to the routing information. An anonymous Internet routing protocol would be a desired approach for this system (e.g. peer-to-peer technology). This needs further research. Topology for anonymous Internet is illustrated in Figure 2.

The anonymous Internet allows the connection between the initiator and the responder to remain anonymous. We call it an anonymous IP tunnel connection. The anonymous connections hide that who is connected to whom, and for what purpose, from both the outside eavesdroppers and compromised AINs. In addition, if the anonymity is also desired at the application layer, all identifying information must be removed from the data stream at the upper layers before being sent over the anonymous IP tunnel connection.

To begin an anonymous session, the initiator sends his/her IP datagrams to the registered AIN using a security connection such as IP ESP+AH tunnel. We call the registered AIN an entrance AIN. The entrance AIN then encrypts the original IP datagrams using the nested encryption described in the Section 5, where the symmetric keys are distributed during the AIVT. According to the destination IP address of the original IP header, the AIN encapsulates the IP datagrams using the IP AH tunnel and sends it to the next AIN if an old AIVT has existed for the source and destination addresses. Otherwise, the AIN creates a new AIVT using a creating AIVT protocol, and then sends the IP datagrams to the next AIN along the AIVT. Final, the AIVT is destroyed by a destroying protocol if the initiator or responder sends a destroying signal after the session, or it is automatically destroyed according to the predefined expiration time.

4.2   Creating PVT Protocol

Based on IETF RFC 2401, the concept of a security association (SA) is fundamental to IPSec. A SA is unidirectional in that it defines the services applied to the IP datagrams transmitted in one direction between the pair of nodes. Both AH and ESP make use of security associations.

To provide secure, bi-directional communication between two AINs, a PVT must comprise four security associations. To make the implementation simple, two of them can use the same parameters except for the destination IP address for IP ESP, and another two for IP AH. An IKE protocol can be used to create the two IP AH tunnels and IP ESP tunnels — a PVT between two AINs.

Note that a PVT is a long-term connection between two AINs, and only provides the signaling transmission services. It doesn't provide the user's data transmission services.

4.3   Creating AIVT Protocol

Every AIVT starts at an entrance AIN, ends with an exit AIN, and passes through several intermediate AINs. The entrance AIN provides the creating function for the AIVT because it is unrealistic to have the user create a suitable route by himself/herself. Actually, sometimes the entrance AIN may only know a part of the whole anonymous Internet topology.

In our anonymous Internet, only the entrance AIN knows the AIVT, and other AIN only knows its previous and next AINs that form the AIVT. Thus, the entrance AIN must be a trusted AIN for the initiator. Our creating AIVT protocol is described as follows.

$$E_{PK_2}(K_2)\ E_{K_2}(\text{AIN}_3,\ \text{Exp-Time},$$
$$E_{PK_3}(K_3)\ E_{K_3}(\text{AIN}_4,\ \text{Exp-Time},$$
$$\ldots\ldots$$
$$E_{PK_{n-1}}(K_{n-1})\ E_{K_{n-1}}(\text{AIN}_n,\ \text{Exp-Time},$$
$$E_{PK_n}(K_n)\ E_{K_n}(\text{NULL},\ \text{Destination IP Address},\ \text{Exp-Time},))\ldots)$$

Figure 1. Creating AIVT Signal.

1. The entrance AIN first identifies a series of AINs forming a route through the anonymous Internet, and constructs a creating AIVT signal according to the source and destination IP addresses of the initiator's IP datagram. An anonymous Internet routing protocol would be required. This will be covered in a future paper. Assuming the route consists of AIN_1, AIN_2, ..., AIN_n where AIN_1 is the entrance AIN and AIN_n is the exit AIN, Figure 1 depicts the creating AIVT signal.
2. The entrance AIN then makes a new random $SPI_{1,2}$, and sends the new $SPI_{1,2}$ and the above creating signal to AIN_2 through the $PVT_{1,2}$ between AIN_1 and AIN_2. AIN_1 and AIN_2 then copies the IP AH of the PVT as an new $TVT_{1,2}$ but using the new $SPI_{1,2}$. AIN_1 stores K_n, K_{n-1}, ..., K_2 as the nested encryption keys for the forward data stream, and the nested decryption keys for the backward data stream.
3. AIN_2 decrypts the creating signal using its private key SK_2, and stores the symmetric key (K_2). AIN_2 then makes a new random $SPI_{2,3}$ and creates a new $TVT_{2,3}$ between AIN_2 and AIN_3 as the step ②. Final, AIN_2 creates a bi-directional connection between $TVT_{1,2}$ and $TVT_{2,3}$, and uses K_2 as the decryption key for the forward data stream and the encryption key for the backward data stream over the connection.
4. All intermediate AINs act as the step ③.
5. Finally, AIN_n decrypts the creating signal using its private key SK_n, and stores K_n as the decryption key for the forward data stream and the encryption key for the backward data stream.

Thus, the AIVT between the entrance AIN and the exit AIN is established using the protocol described above for the initiator and responder. Since the data confidentiality is supported by the nested encryption operation at the entrance AIN, IP AH is enough to provide the secure protection for the TVT. Figure 2 depicts an AIVT.

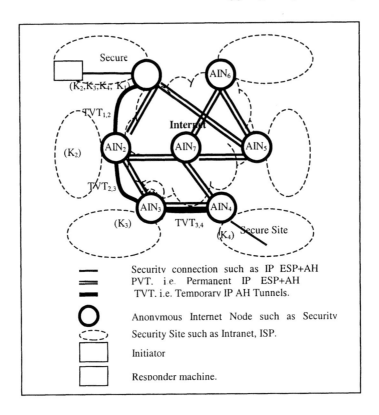

Figure 2. Anonymous IP-datagram Virtual Tunnel.

4.4 Anonymous End-to-end Data Transmission Protocol

In our architecture, we assume that there is an IP ESP+AH tunnel between the initiator and the entrance AIN, and also between the exit AIN and the responder. The advantages are as follows. First, the entrance AIN and exit AIN can support the data origin authentication for the initiator and responder, respectively. On the other hand, it can hide the responder's IP address from the outside eavesdroppers between the initiator and the entrance AIN, and the initiator's IP address from the outside eavesdroppers between the exit AIN and the responder.

In addition, based on whether the initiator hides its IP address from the exit AIN and the responder or not, two options are available for providing different degree anonymous bi-directional data transmission. The initiator can choose the best one according to its requirements for privacy and security.

Anonymous Internet Communication Based on IPSec

The first option is that the initiator hides its IP address from any entities except for the entrance AIN. In this situation, the initiator can get higher degree anonymity protection. The second option is that the initiator doesn't hide its IP address from the exit AIN and the responder. In this situation, the advantage is that it also can support some end-to-end applications at IP layer, for example an anonymous end-to-end IPSec and VPN, etc. The difference is that the initiator chooses a random private IP address as the source address of the original IP-datagram in the first situation, and a real IP address as the source address in the second situation. The technique is described as follows.

1. The initiator first prepares his/her original IP-datagram according to his/her option for the anonymous degrees, i.e. using a random private IP address or real IP address as the source IP address of the original IP datagrams, and then sends the original IP datagrams to the entrance AIN using an IP ESP+AH tunnel ($PVT_{i,1}$) between the initiator and the entrance AIN.
2. The entrance AIN verifies whether the initiator is its registration customer using IP AH and gets the real IP source address, and then decrypts the data payload using IP ESP and gets the original IP header. The entrance AIN then creates an $AIVT_{1,2}$ using the above creating protocol according to the real source address and original IP header, makes a connection between the $PVT_{i,1}$ and the $AIVT_{1,2}$, and encrypts the whole IP datagram using the nested encryption technique, i.e. the IP datagram first is encrypted with K_n, then K_{n-1}, ..., and final K_2. Final, the entrance AIN creates a new IP AH header for the encrypted data payload according to the $TVT_{1,2}$ and puts the new IP datagram to the $TVT_{1,2}$.
3. AIN_2 verifies the IP AH and decrypts one layer of the encryption data payload using its decryption key K_2, and then creates a new IP AH header according to the $TVT_{2,3}$, and sends the new IP datagram to the next AIN.
4. All intermediate AINs act as the step ③.
5. Final, AIN_n verifies the IP AH, decrypts last layer encryption using its decryption key K_n, and gets the original IP datagram. The exit AIN then sends the original IP datagram to the responder using the IP ESP+AH tunnel ($PVT_{n,r}$) between the exit AIN and the responder. In the meantime, AIN_n makes a connection between the $PVT_{n,r}$ and the $AIVT_{n1,n}$.

When a backward data stream is sent from the responder, an inverse processing is performed as the above, but the cryptographic operation is an encryption for each node except for the entrance AIN. The entrance AIN decrypts the backward data stream with K_2, then K_3,..., and final K_n.

4.5 Destroying AIVT Protocol

Every AIVT has a default inactive expiration time. It may also have another expiration time. An AIVT will be destroyed immediately after a destroying signal is sent, or when its expiration time or the default inactive expiration time is expired. A destroying AIVT signal can be made and sent by the initiator, responder and any AINs in the AIVT. There are several situations for destroying an AIVT as follows.

The first situation is for the expiration time and default inactive expiration time. If the AIVT has an expiration time and it has expired, all AINs automatically destroy the AIVT according to the expiration time. If the AIVT doesn't have an expiration time, all AINs also automatically destroy the AIVT according to its default inactive expiration time.

The second situation is for the initiator or responder sending a destroying signal. If either the initiator or responder sends a destroying signal to the entrance AIN or exit AIN for any reason (e.g. a session ends), the entrance AIN or exit AIN will create a destroying AIVT signal using the PVTs, send it to the next AIN, and destroy the AIVT. The next AIN finds its next SPI, and then creates a new destroying AIVT signal and sends it to its next AIN, and so on. Figure 3 depicts the AIVT destroying signal.

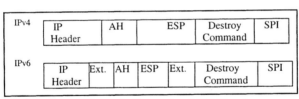

Figure 3. Destroying AIVT Signal.

The final situation is for any AINs sending a destroying AIVT signal. If any AIN sends a destroying AIVT signal for some reasons (e.g. it cannot connect the next AIN), the AIN will create two destroying AIVT signals, send them to the next AIN along the two directions using the PVTs, and destroy the AIVT. The destroying AIVT signal is the same as Figure 5.

5 IMPLEMENTATION

The easiest way to build the anonymous Internet without requiring the complete redesign and deployment of new client and server software is to make use of existing IPSec software technologies.

This section presents the interface specification between the components in the anonymous Internet. In order to provide some structure to this specification, we discuss the components in the order that data would move from the initiator to the responder in this section.

5.1 Entrance AINs

The interface between the initiator and the entrance AIN is defined as a standard IP ESP+AH tunnel, and is independent of any special applications. The initiator sends any IP datagrams to the entrance AIN using the IP ESP+AH tunnel. Figure 4 depicts the IP ESP+AH tunnel format.

IPv4	New IP Header		AH	ESP	Original IP Header		Upper layer Data payload
IPv6	New IP Header	Ext.	AH	SP	Original IP Header	Ext.	Upper layer Data payload

Figure 4. IP ESP+AH Tunnel between the Initiator and the Entrance

The initiator and the entrance AIN have four SAs for the bi-directional IP ESP+AH tunnel, respectively. Two of them support the forward data stream for IP ESP and IP AH, respectively. Another two support the backward data stream. The IP ESP+AH tunnel is built by the IKE protocol. Of course, they also can only use IP ESP or AH as the IP tunnel, but it would weaken the security.

The interfaces between the entrance AIN and next AIN include the PVT interface and the TVT interface. The PVT interface is defined as a standard IP ESP+AH tunnel. It is a long-term tunnel that can be built by the IKE protocol previously. Figure 5 depicts the PVT format for supporting the signaling transmission services.

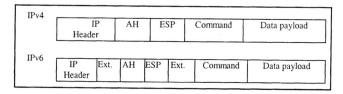

Figure 5. PVT Signaling Format.

The Command and Data payload fields are always encrypted using the IP ESP. The command is CREATE (0), DESTROY (1), or other. If the command is CREATE, the data payload should be *SPI+Creating Signal*. If the command is DESTROY, the data payload should be *SPI+Padding*.

The TVT interface is defined as a standard IP AH tunnel. It is a temporary tunnel that is built by the creating AIVT protocol. Figure 6 depicts the TVT format for supporting the data transmission services.

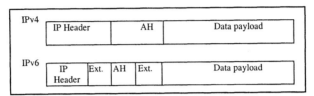

Figure 6. TVT Data Transmission

The above data payload is encrypted with a nested encryption operation. Figure 7 depicts the nested encryption data through the TVT. For the forward data stream, the data payload is created by the entrance AIN using $K_n, K_{n-1}, ..., K_2$ to encrypt the original IP datagram. For the backward data stream, the data payload is created by the exit AIN and intermediate AINs using their keys to encrypt the original IP datagram.

$$E_{K_2}(E_{K_3}(...E_{K_{n-1}}(E_{K_n}(\text{Original IP Datagram}))...)$$

Figure 7. Nested Encryption for Forward Data.

At the entrance AIN, a routing table is built for the AIVT according to the real IP source address, the original IP header of the initiator's IP datagram and the TVT. Two route messages are added to the routing table when the AIVT is built. Thus, the IP datagrams from the initiator to the responder are sent to the $TVT_{1,2}$, and the IP datagrams from the $TVT_{1,2}$ are sent to the initiator. Table 1 depicts the routing table in the entrance AIN.

Table 1. Routing Table in the Entrance

Source	Destination
$(IP_I, IP_R)_{Tunnel}$	$TVT_{1,2}$
$TVT_{1,2}$	$(IP_R, IP_I)_{Tunnel}$
...	...

Where $(IP_I, IP_R)_{Tunnel}$ means an IP ESP+AH tunnel with the original IP header from the initiator to the entrance AIN. The IP_I is the source IP address in the original IP header and may be real or unreal, and the IP_R is the responder's IP address.

5.2 Intermediate AINs

The interfaces between two neighboring intermediate AINs are the same as the interfaces between the entrance AIN and the next AIN, i.e. two interfaces: one for the PVT, and another for the TVT.

In an intermediate AIN, a routing table is built for the AIVT according to the its previous TVT and its next TVT in the AIN. Two route messages are added to the routing table when the AIVT is built. For example in the AIN_2, the IP datagrams are sent from the $TVT_{1,2}$ to the $TVT_{2,3}$ for the forward IP datagrams, and from the $TVT_{2,3}$ to the $TVT_{1,2}$ for the backward IP datagrams. Table 2 depicts the routing table in the AIN_2.

Table 2. Routing Table in the AIN_2

Source	Destination
$TVT_{1,2}$	$TVT_{2,3}$
$TVT_{2,3}$	$TVT_{1,2}$
...	...

In addition, each intermediate AIN only decrypts one layer of the forward data stream and encrypts one layer of the backward data stream, and sends it to the next AIN by the routing table.

5.3 Exit AINs

The interfaces between the next-to-last AIN and the exit AIN also are the same as the interfaces between the entrance AIN and the next AIN, i.e. two interfaces: one for the PVT, and another for TVT.

IP ESP+AH also is the desired interface between the exit AIN and the responder. The routing table is similar to the routing table of the entrance AIN. But the exit AIN is not sure whether the source IP address of the original IP datagrams is real or not. Table 3 depicts the routing messages added to the routing table it the exit AIN.

Table 3. Routing Table in the Exit

Source	Destination
$TVT_{n-1,n}$	$(IP_I, IP_R)_{Tunnel}$
$(IP_R, IP_I)_{Tunnel}$	$TVT_{n-1,n}$
...	...

Final, the exit AIN decrypts the last layer of the encryption data using its decryption key K_n for the forward data stream and sends it to the responder. For the backward data stream, the exit AIN encrypts the original IP datagram using K_n and sends it to the AIN_{n-1}.

5.4 Other Consideration

In order to provide ideal personal data protection against traffic analysis, the following techniques need to be included in our system.

- *Packet Padding:* In order to protect against the packet size correlation attack, the packet padding technique must be used in our system to make each packet transmitted between any two nodes (including the initiator and responder) have the same size.
- *Dummy Traffic:* In order to protect against the packet timing correlation attack, the dummy traffic technique must be used to send a constant number of packets per unit time.
- *Ticket:* In order to protect against the flooding attack, the ticket technique can be added to our system, i.e. each user must show that he/she is allowed to use the system at the respective time slice by providing a ticket valid for the certain slice, only.

Finally, in order to be effective for our system, every entrance and exit AIN also take part in the role of the intermediate AINs.

6 VULNERABILITIES

Our Anonymous Internet Communication provides personal data protection against traffic analysis at IP layer. It can prevent the following attacks:

- *Collusion Attack:* It can prevent the collusion attack among any intermediate AINs and outside eavesdroppers because every intermediate AIN only knows its previous and next AINs. But it does not provide the protection against the attacks from the entrance AIN.
- *Message Coding Attack:* It provides the protection against the message coding attack using the nested encryption.
- *Message Volume Attack:* It can prevent the message volume attack using the packet padding technique.
- *Timing Attack:* It can prevent the timing attack using the dummy traffic techniques [1].

- **Flooding Attack:** It can prevent the flooding attack using the "ticket" technique [1].
- **Replay attack:** It provides the protection against the replay attack using IP AH, and data origin authentication and integrity protection. It also provides the data confidentiality protection with IP ESP for the signaling data and with the nested encryption for the transmission data.

In our system, the entrance AIN retains all information necessary for the anonymous connection such as the initiator, the responder and all AINs. Thus, the entrance AIN must be a trusted node by the initiator. In addition, any outside eavesdroppers between the AINs would not be able to discover anything about the data transfer.

7 CONCLUSION

IPSec is an excellent approach for secure communication over the Internet. Based on IPSec, we present an anonymous Internet communication using IP tunnel at IP layer. It provides bi-directional, real-time communication for any TCP and UDP applications, including the end-to-end IPSec, VPN, SSL, etc.

In this paper, however we don't discuss the network information database, routing and how to learn the topology of the anonymous Internet. This will be the subject of a future paper.

Because of its importance to the acceptance of e-business systems, we expect that more and more people will be involved in the research and development of an anonymous Internet including pseudonym IP. All of these efforts will make our Internet more robust and secure.

8 ACKNOWLEDGMENTS

We would like to thank all members of IIT at the NRC of Canada for their support towards our R&D projects in Privacy Protection. We also thank our European Community partners in the Privacy Incorporated Software Agent (PISA), an IST-EU Fifth Framework Project.

9. REFERENCES

[1] O.Berthold, H.Federrath and S.Kopsell. Web MIXes: A System for Anonymous and Unobservable Internet Access. In H.Federrath, editor, Anonymity 2000, Volume 2009 of Lecture Notes in Computer Science, pages 115-129, Springer-Verlag, 2000.
[2] O.Berthold, H.Federrath and M.Kohntopp. Project "Anonymity and Unobservability" in the Internet. Proceedings of the tenth conference on Computers, freedom and privacy: challenging the assumptions, Toronto, ON Canada, pages 57-68, April 2000.
[3] P.Boucher, A.Shostack and I.Goldberg. Freedom Systems 2.0 Archtecture. December 2000. Available at http://www.freedom.net/info/whitepapers/Freedom_System_2_Architecture.pdf.
[4] D.Chaum. Untraceable Electronic Mail, Return Address, and Digital Pseudonyms. Communications of the ACM, Vol.24, No.2, pages 84-88, 1981.
[5] W.Dai. Pipenet 1.1. 2000. Available at http://www.eskimo.com /~weidai/pipenet.txt.
[6] A.O.Freier, P.Karlton and P.C.Kocher. The SSL Protocol: Version 3.0. Internet Draft, Netscape Communications, March 1996. Available at http://home.netscape.com/eng/ssl3/ssl-toc.html.
[7] D.Goldschlag, M.Reed and P.Syverson. Hiding Routing Information. In R.Anderson, editor, Information Hiding: First International Workshop, Volume 1174 of Lecture Notes in Computer Science, pages 137-150, Springer-Verlag, 1996.
[8] S.Kent and R.Atkinson. Security Architecture for the Internet Protocol. IETF RFC 2401, November 1998.
[9] S.Kent and R.Atkinson. IP Authentication Header. IETF RFC 2402, November 1998.
[10] S.Kent and R.Atkinson. IP Encapsulating Security Payload (ESP). IETF RFC 2406, November 1998.
[11] M.Reed, P.Syverson and D.Goldschlag. Anonymous Connections and Onion Routing. IEEE Journal on Selected Areas in Communications, Vol.16, No.4, pages 482-494, May 1998.
[12] M.Reiter and A.Rubin. Crowds: Anonymity for Web Transactions. ACM Transactions on Information and System Security, Vol.1, pages 66-92, 1998.

Internet Interconnection Economic Model and its Analysis: Peering and Settlement

Martin B. Weiss and SeungJae Shin
Dept. of Info. Science and Telecom.
School of Information Science
University of Pittsburgh
*(*mbw@pitt.edu *and* sjshin@mail.sis.pitt.edu*)*

Abstract: Peering and transit are two types of Internet interconnection among ISPs. Peering has been a core concept to sustain Internet industry. However, for the past several years, many ISPs broke their peering arrangement because of asymmetric traffic pattern and asymmetric benefit and cost from the peering. Even though traffic flows are not a good indicator of the relative benefit of an Internet interconnection between the ISPs, it is needless to say that cost is a function of traffic and the only thing that we can know for certain is inbound/outbound traffic volumes between the ISPs. In this context, we suggest Max {inbound traffic volume, outbound traffic volume} as an alternative criterion to determine the Internet settlement between ISPs and we demonstrate this rule makes ISPs easier to make a peering arrangement. In our model, the traffic volume is a function of a market share. We will show the market share decides traffic volume, which is based on the settlement between ISPs. As a result, we address the current interconnection settlement problem with knowledge of inbound and outbound traffic flows and we develop an analytical framework to explain the Internet interconnection settlement.

Keywords: Internet interconnection, peering, transit, settlement

1 INTRODUCTION

The Internet industry is dynamic. The number of Internet Service Providers (ISPs) is increasing rapidly and the structure of the industry changes continuously. It is widely accepted that today's Internet industry has vertical structure: over 40 Internet Backbone Providers (IBPs) including

5 top-tier backbones constitute the upstream industry (Kende, 2000) and over 10,000 ISPs for accessing the Internet make up the downstream industry (Weinberg, 2000). A backbone provider service is critical for those ISPs to connect to the whole Internet. As the number of ISPs and IBPs increase, the Internet interconnection settlement[1] issue is becoming more significant. Under the current interconnection arrangement, it is uncertain to decide which (a sender or a receiver) has responsibility for the traffic to send or receive because current capacity based interconnection pricing scheme does not know which part has to pay a cost of that traffic. ISPs can use pricing based on inbound traffic volume, on outbound traffic volume, on a hybrid of inbound and outbound traffic volume, or on the line capacity regardless of volume (Hutson, 1998, p565). None of these methods gives full satisfaction to all of the service providers in the industry. Many scholars and industry experts say that a usage-based pricing scheme and a usage-based settlement system are the only alternative to overcome the current uncertainty. However, they agree that there are technical difficulties in changing the current Internet financial system to a usage-based system[2]. Even though traffic flows are not a good indicator of the relative benefit of an Internet interconnection between the service providers (Kende, 2000, p36), it is needless to say that cost is a function of traffic and the only thing that we can know for certain is inbound/outbound traffic volumes between the service providers. We address the current interconnection settlement problem with knowledge of inbound and outbound traffic flows and we develop an analytical framework to explain the Internet interconnection settlement issues.

2 TYPES OF INTERNET INTERCONNECTION

There are two types of Internet interconnection among ISPs and IBPs: peering and transit. The only difference among theses types is in the financial rights and obligation that they generate to their customers. First, we will examine the history of Internet interconnection.

2.1 History of Interconnection

To understand the relationship between peering and transit, it is necessary to review the situation before the commercialization of the

---

[1] Settlement can be thought of as payments or financial transfers between ISPs in return for interconnection and interoperability (Cawley, 1997)
[2] Various kinds of protocols, duplicate layers, packet dropping, dynamic routing, and reliability issues, etc.

Internet in 1995. During the early development of the Internet, there was only one backbone and only one customer, the military, so interconnection was not an issue. In the 1980s, as the Internet was opened to academic and research institutions, and the National Science Foundation (NSF) funded the NSFNET as an Internet backbone. Around that time, the Federal Internet Exchange (FIX) served as a first point of interconnection between federal and academic networks. At the time that commercial networks began appearing, general commercial activity on the Internet was restricted by Acceptable Use Policy (AUP), which prevented the commercial networks from exchange traffic with one another using the NSFNET as the backbone. (Kende, 2000, p5) In the early 1990s, a number of commercial backbone operators including PSINet, UUNET, and CerfNET established the Commercial Internet Exchange (CIX) for the purpose of interconnecting theses backbones and exchanging their end users' traffic. The NSF decided to cease to operate the NSF backbone, which was replaced by four Network Access Points (NAPs)[3]. (Minoli and Schmidt, 1998, p28) The role of NAPs is similar to that of CIX. After the advent of CIX and NAPs, commercial backbones developed a system of interconnection known as peering.

2.2 What is Peering?

The term "peering" is sometimes used generically to refer to Internet interconnection with no financial settlement, which is known as "Sender Keeps All (SKA)" or "Bill and Keep" arrangement. Peering can be divided into several categories: (1) according to its openness, it can be private peering or public peering, (2) according to the numbers of peering partners it can be Bilateral Peering Arrangement (BLPA) or Multilateral Peering Arrangement (MLPA), and (3) according to the market in which it occurs, it can be primary peering in the backbone market or secondary peering in the downstream market.

The original 4 NAPs points for public peering. Anyone who is a member of NAP can exchange traffic based on equal cost sharing. Members pay for their own router to connect to the NAP plus the connectivity fee charged by the NAP. As the Internet traffic grew, the NAPs suffered from congestion. Therefore, direct circuit interconnection between two large IBPs was introduced, so called bilateral private peering, which takes place at a mutually agreed place of interconnection. This private peering is opposed to public peering that takes place at the NAPs. It is estimated that 80 percent of Internet traffic is exchanged via private peering (Kende, 2000, pp. 6-7).

---

[3] 4 NAPs are Chicago NAP (Ameritech), San Francisco NAP (Pacific Bell), New York NAP (Sprint), and Washington D.C. NAP (Metropolitan Fiber Systems).

A peering arrangement is based on equality, that is, ISPs of equal size would peer. The measures of size could be (i) geographic coverage, (ii) network capacity, (iii) traffic volume, (iv) size of customer base, or (v) a position in the market. The ISPs would peer when they perceive equal benefit from peering based on their own subjective terms. (Kende, 2000, p8) The followings are peering characteristics:

1. Peering partners only exchange traffic that originates with the customer of one ISP and terminates with the customer of the other peered ISP. As part of peering arrangement, an ISP would not act as an intermediary. And it would not accept the traffic of one peering partner for the purpose of transiting this traffic to another peering partner. This characteristic is called a "non-transitive relationship."
2. Peering partners exchange traffic on a settlement-free basis. The only cost of each partner is its own equipment and the transmission capacity needed for the two peers to meet at each peering point
3. Peering partners generally meet in a number of geographically dispersed locations. In order to decide where to pass traffic to another, they have adopted what is known as "hot-potato routing," where an ISP will pass traffic to another backbone at the earliest point of exchange.

According to Block (Cukier, 1999), there are two conditions necessary SKA peering, that is, peering with no settlement, to be viable: (1) The traffic flows should be roughly balanced between interconnecting networks; and (2) the cost of terminating traffic should be low in relation to the cost of measuring and billing for traffic. In sum, peering is sustainable under the assumption of mutual benefits and costly, unnecessary traffic measuring. Peering partners would make a peering arrangement if they each perceive that they have more benefits than costs from the peering arrangement. Most ISPs historically have not metered traffic flows and accordingly have not erected a pricing mechanism based on usage. Unlimited access with a flat rate is a general form of pricing structure in the Internet industry. Finally, peering makes billing simple: no metering and no financial settlement.

2.3 What is Transit?

Transit is an alternative arrangement between ISPs and IBPs, in which one pays another to deliver traffic between its customers and the customers of other provider. The relationship of transit arrangement is hierarchical: a provider-customer relationship. Unlike a peering relationship, a transit provider will route traffic from the transit customer to its peering partners. An IBP with many transit customers has a better position when negotiating a peering arrangement with other IBPs. Another difference between peering and transit is existence of Service Level Agreement (SLA). In the peering arrangement, there is no SLA to guarantee rapid repair of problems. In the case of an outage, both peering partners may try to repair the problem, but it is not mandatory. This is one of the reasons peering agreements with the company short of competent technical staffs are broken. But in the transit arrangement it is a contract and customers could ask transit provider to meet the SLA. In the case of e-commerce companies, they prefer transit to peering. Since one minute of outage causes lots of losses to them, rapid recovery is critical to their business. Furthermore, in the case of transit, there is no threat to quit the relationship while in the case of peering a non-renewal of the peering agreement is a threat. When purchasing transit service, ISPs will consider other factors beside low cost: performance of the transit provider's backbone, location of access nodes, number of directly connected customers, and a market position.

3 ASSUMPTIONS OF THE INTERNET INTERCONNECTION ECONOMIC MODEL

To simplify the model, we made several assumptions:

1. There is only one IBP in the upstream backbone market and there are two ISPs in the downstream Internet access market. ISP-1 and ISP-2 sell the Internet connectivity to their customers. The interconnection to the IBP's network is the only way to do that because the IBP is the sole provider of the Internet backbone. The IBP does not support customers directly.
2. If the market share of ISP-1 is α, then that of ISP-2 is $(1-\alpha)$ because there are only two ISPs in the market ($0<\alpha<1$). This market share is a unique factor in determining traffic volume and α is a parameter, which can be given by the downstream Internet access market.
3. N is the total number of customers in the downstream Internet access market. If the market share of ISP-1 is α, the number of Internet subscribers of ISP-1 is $N*\alpha$ and that of ISP-2 is $N*(1-\alpha)$. Customers do not have to choose two ISPs at the same time because of homogeneity of the service.
4. There are three kinds of traffic in each ISP, T_{ij}, where i = 1, 2 (subscript for ISP) and j = L, O, I (subscript for traffic type): local traffic (T_{1L}, T_{2L}), outbound traffic (T_{1O}, T_{2O}), and inbound traffic (T_{1I}, T_{2I}).
   - Each ISP generates two types of traffic: local traffic which uses only local network and outbound traffic which uses whole network including backbone network and the other ISP's network.

- The amount of local traffic depends on its market share. An ISP with large market share has a relatively large portion of local traffic compared to its outbound traffic, while an ISP with small market share has a relatively small portion of local traffic compared to its outbound traffic.
- The amount of outbound traffic depends on the other ISP's market share. That is to say, ISP-1's outbound traffic (T_{1O}) depends on the market share of ISP-2, $(1-\alpha)$, and vice versa.
- One ISP's inbound traffic is the same as the other ISP's outbound traffic, which means ISP-1's inbound traffic (T_{1I}) is the same as ISP-2's outbound traffic (T_{2O}) because there are only two ISPs in the market.
- The assumption of dependency of market share comes from a concept of network externality[4]. An ISP with large customer base and many Internet resources such as FTP sites and Web sites is less dependent on other ISPs than an ISP with small customer base and less Internet resources.

5. The average traffic generated per subscriber is assumed that 3 G bits per month, which comes from the following assumptions and calculation:
   - The two ISPs sell only 56 Kbps dial-up modem Internet connectivity[5].
   - The average hours of Internet usage is 90 hours per month[6].
   - We apply 1:6 bandwidth ratio. The bandwidth ratio occurs because a user does not consume whole 56 Kbps for the duration of the connection. For example, in the case of reading a news article in the Internet, a user can use a full capacity of downloading that news article, but after that, when he reads a news article, no traffic is transmitted.
   - 56 (Kbps) * 90 (hours/month) * 3600 (seconds/hour) * 1/6 = 3 G bits per subscriber approximately.

6. The number of subscriber and the average traffic per subscriber determine the traffic volumes such as T_1 and T_2.

---

[4] Network externalities arise when the value or utility that a customer derives from a product or service increases as a function of other customers of the same or compatible products or services; that is, the more users there are, the more valuable the network. There are two kinds of network externality in the Internet. One is direct network externality: the more E-mail users, the more valuable the Internet is. The other is indirect network externality: the more Internet users there are, the more web contents will be developed, which makes the Internet even more valuable for its users. (Kende, 2000)

[5] According to the report (2001) from USGAO (United States Government Account Office), 87.5% of Internet users use the narrow band dial-up modem.

[6] According to the report (2001) from USGAO (United States Government Account Office), 15 ~ 25 hours per week is the most frequently selected Internet usage hours from which we choose 20 hours per week and converted it to 90 hours per month.

- The total traffic generated by ISP-1 (T_1) is the sum of local traffic (T_{1L}) and outbound traffic (T_{1O}): $T_1 = T_{1L} + T_{1O}$ = 3 G bits*(Number of ISP-1's Subscriber) = 3 G bits *(α*N).
- The total traffic generated by ISP-2 (T_2) is the sum of local traffic (T_{2L}) and outbound traffic (T_{2O}): $T_2 = T_{2L} + T_{2O}$ = 3 G bits *(Number of ISP-2's Subscriber) = 3 G bits *(1-α)*N.

7. The interconnection fee (settlement) is the only cost of each ISP to IBP. The IBP does not have any cost. This settlement is the only revenue of the IBP.
8. Each ISP gets a fixed monthly access charge from its own users.

The following graph illustrates traffic flow and relationship between IBP and ISPs.

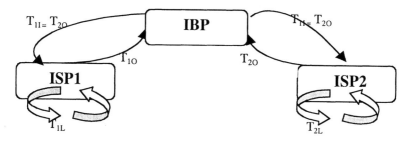

Figure 1. Traffic flows

4   TRAFFIC FUNCTION

Because there are only two ISPs, if one ISP's market share is 100%, then all the traffic generated in this market comes from that ISP and therefore no connection to the backbone is needed. If the market shares of the two ISPs are the same (α=50%), the traffic from each ISP must be the same, because the traffic is dependent upon only the market share. The local traffic increases and the outbound traffic decreases as the market share increases, but the increasing or decreasing patterns are not linear. This non-linear characteristic comes from the number of servers on a network. Under the assumption that each user accesses each server with equal probability, average traffic is directly proportional to the number of servers on the network. The local traffic function (f_L) is non-linear only if the number of servers on the network is non-linear with respect to market share (α). One of the main reasons for non-linear number of servers with respect to market share (α) is mirroring[7] and caching[8]. The mirroring of servers like duplicate FTP servers or the caching of Web pages by ISPs, moves applications and data closer to the Internet users. As the number of subscribers increases, the mirroring servers and the caching servers increase more than linear, which means that local traffic also increases in a non-linear fashion. According to the report by Williams and Sochats (1999), investigating whether the amount of time Internet applications hold Interstate telecommunication resources is less than 10% of the total time an Internet user is connected to its local ISP, mirrored sites continue to expand and caching of Web pages by ISPs even more widespread, the percentage of time Internet applications will use the Interstate transmission facilities will continue to decrease. This result implies that as the number of the mirrored and caching servers increases, the portion of local traffic within the network increases sharply.

By use of the above concept, we use the quadratic monotonic increasing function satisfying the following three points in the market share plane (X-axis) and traffic plane (Y-axis): (0%, 0%), (50%, 50%), and (100%, 100%). The local traffic function (f_L) and outbound traffic function (f_o) satisfying the above conditions are:

$$f_L(\alpha) = \begin{cases} 2*\alpha^2 & : 0<\alpha<0.5 \quad \text{...............(1)} \\ -2*(1-\alpha)^2 +1 & : 0.5<\alpha<1 \quad \text{...............(2)} \end{cases}$$

$$f_o(1-\alpha) = 1 - f_L(\alpha) = \begin{cases} 1 - 2*\alpha^2 & : 0<\alpha<0.5 \quad \text{...............(3)} \\ -2*(1-\alpha)^2 & : 0.5<\alpha<1 \quad \text{...............(4)} \end{cases}$$

The following figure shows graphs of local and outbound traffic using the above functions.

---

[7] Content providers maintain duplicated web sites on a number of different servers to improve bandwidth efficiency and better service to the end-users.
[8] Storing copies of frequently retrieved Web pages or data on servers that are geographically closer to end-users.

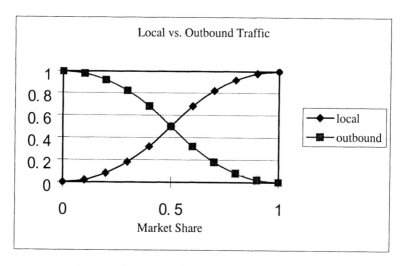

Figure2. Proportion of local and outbound traffic

5 TRAFFIC SUPPLY FOR INTERCONNECTION

As we mentioned earlier section, the traffic using ISP-1's network (T_{1N}) consists of three types of traffic: (1) Local traffic only using ISP-1's network (T_{1L}), (2) outbound traffic generated by ISP-1 and forwarding to ISP-2's network (T_{1O}), and (3) inbound traffic generated by ISP-2 and coming into ISP-1's network (T_{1I}). For the traffic using ISP-1's network and ISP-2's network, we can write the following equations:

$$T_{1N} = T_{1L} + T_{1O} + T_{1I} \quad \ldots(5)$$
$$T_{2N} = T_{2L} + T_{2O} + T_{2I} \quad \ldots(6)$$

One ISP's outbound traffic is the same as the other's inbound traffic. Therefore we can change equations (5) & (6) into equations (7) & (8):

$$T_{1N} = T_{1L} + T_{1O} + T_{2O} \quad \ldots(7)$$
$$T_{2N} = T_{2L} + T_{2O} + T_{1O} \quad \ldots(8)$$

According to equations (7) & (8), the traffic of one ISP's network is the sum of its local traffic and total outbound traffic ($T_{1O} + T_{2O}$). The line capacity between ISP and IBP should be at least the sum of each ISP's outbound traffic, which can be traffic supply in this model. We assume that the IBP has enough capacity to accommodate ISPs' traffic in the downstream market.

6. MAXIMUM OUTBOUND TRAFFIC CRITERION FOR SETTLEMENT AND TRAFFIC DEMAND

These days, 80%-85% of all ISP traffic is traffic generated by receiver requests, such as Web page retrieval or downloading files (Hutson, 1998, p571). In these cases, the request traffic is generally very small compared to the traffic of web pages or downloaded files. That means there is much more inbound traffic compared to the outbound traffic for each service request. This is a typical asymmetric traffic flow pattern between inbound and outbound traffic.

The peering arrangement is based upon the symmetric traffic pattern. However, service oriented Internet providers such as Web hosting providers (web farm) produce more outbound traffic than the providers with a large customer base. Peering is basically SKA settlement system. Refusing a peering arrangement means that the refused partner has to pay the transit fee for interconnection service. The inbound and outbound traffic volumes are not a good indicator to determine Internet connectivity fees because we do not know well who can get more benefit and who uses more the other's network precisely. There is an uncertainty in matching benefit and cost of network usage. However, in reality, the knowledge of inbound and outbound traffic volumes is the only available information between ISPs' networks. Regardless of the type of service provider and the customer size of ISP, when an ISP connects to a backbone provider, **Max {inbound traffic volume, outbound traffic volume}** can be an alternative criterion to determine the interconnection fee between the two providers.

In the case of interconnection between ISP-1 and IBP, Max {inbound traffic volume, outbound traffic volume} can be rewritten as,

$$\text{Max } \{T_{10}, T_{11}\}$$
$$= \text{Max } \{T_{10}, T_{20}\}$$
$$= \text{Max } \{f_O(\alpha)*T_1, f_O(1-\alpha)*T_2\}$$
$$= \text{Max } \{(1-f_L(\alpha))*T_1, (1-f_L(1-\alpha))*T_2\}$$
$$= \text{Max } \{(1-f_L(\alpha))*\alpha*3G \text{ bits}*N, (1-f_L(1-\alpha))*(1-\alpha)*3G \text{ bits}*N\}$$
$$= \begin{cases} (-2\alpha^3 + \alpha) * 3G \text{ bits }*N, \text{ when } 0 < \alpha < 0.5 \quad \dots(9) \\ 2*(1-\alpha)^3 * 3G \text{ bits}*N, \text{ when } 0.5 < \alpha < 1 \quad \dots(10) \end{cases}$$

The equations (9) & (10) consist of two parts: deterministic terms (3 G bits *N) and variable terms that can be varied by the value of market share (α). The following table and graph are made by the only variable part of the equations (9) and (10) when a market share of ISP-1 (α) increases from 0% to 100%.

Internet Interconnection Economic Model and Its Analysis

Market Share		Local Traffic		Outbound Traffic		Inbound Traffic		Max
ISP-1	ISP-2	ISP-1	ISP-2	ISP-1	ISP-2	ISP-1	ISP-2	{T_{10}, T_{20}}
α	$(1-\alpha)$	T_{1L}	T_{2L}	T_{10}	T_{20}	T_{11}	T_{21}	
0.00	1.00	0.000	1.000	0.000	0.000	0.000	0.000	0.000
0.10	0.90	0.020	0.980	0.980	0.020	0.020	0.980	0.098
0.20	0.80	0.080	0.920	0.920	0.080	0.080	0.920	0.184
0.30	0.70	0.180	0.820	0.820	0.180	0.180	0.820	0.246
0.40	0.60	0.320	0.680	0.680	0.320	0.320	0.680	0.272
0.50	0.50	0.500	0.500	0.500	0.500	0.500	0.500	0.250
0.60	0.40	0.680	0.320	0.320	0.680	0.680	0.320	0.272
0.70	0.30	0.820	0.180	0.180	0.820	0.820	0.180	0.246
0.80	0.20	0.920	0.080	0.080	0.920	0.920	0.080	0.184
0.90	0.10	0.980	0.020	0.020	0.980	0.980	0.020	0.098
1.00	0.00	1.000	0.000	0.000	0.000	0.000	0.000	0.000

Table 1. Max {{T_{10}, T_{20}}

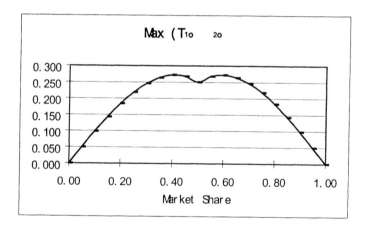

Figure 3. Max {T_{10}, T_{20}}

The maximum outbound traffic of the ISP-1 and ISP-2 can be the traffic demand for each ISP to determine settlement in this model.

- Traffic Demand (α) = Max {T_{10}, T_{20}}

From the above table and graph, we can draw the conclusion that the traffic demand for interconnection with the backbone provider depends on the market share α. And this demand is symmetric at 50%. The following table shows the traffic demand per month according to the market share α.

Share α	0.1 / 0.9	0.2 / 0.8	0.3 / 0.7	0.4 / 0.6	0.5
Traffic Demand	294Mbits*N (=3G*0.098)	552Mbits *N (=3G*0.184)	738Mbits *N (=3G*0.246)	816Mbits *N (=3G*0.272)	750Mbits N (=3G*0.250)

Table 2. Traffic Demand

The number of bits that can be accommodated by a single T-1 line in a month is 432,000 M bits per month[9], which is calculated by the assumption of 4 peak hours a day. We can know how many T-1 lines are needed if we divide traffic demand per month by 432, 000 M bits. The following table shows the number of T-1 lines needed for traffic demand under the assumption of N = 5,000 users. For example, in the case of α = 0.1, {294 M bits * 5,000}/ 432,000 M bits = 3.4 T-1 lines.

Market Share (α)	0.1 / 0.9	0.2 / 0.8	0.3 / 0.7	0.4 / 0.6	0.5
Number of T-1s	4	7	9	10	9

Table 3. Number of T-1s needed for Traffic Demand

The settlement is generally defined as a product of the traffic demand and the interconnection fee. In this case, the settlement can be expressed as a product of (Number of T-1 lines) and (T-1 transit price). For example, if the ISP-1 uses 4 T-1 lines and the Internet interconnection T-1 transit price is $1,000 per month, then the settlement is $4,000 per month. The transit price is usually determined by the provider's relative strength and level of investment in a particular area (Halabi, 2000, 42p). It is certain that T-1 transit prices have decreased continuously. In 1996, the Internet connectivity for T-1 was $3,000 per month with $1000 setup fee (Halabi, 1997, 40p). According to Martin (2001), the average price of a T-1 connection has continuously decreased for several years: $1,729 (1999), $1,348 (2000), and $1,228 (2001).

We assume that the T-1 transit price is given by the market, which is equal to $1,000 per month. The following table shows the amount of settlement payment from each ISP to the IBP. This settlement is also symmetric at the mid point of α= 0.5.

Market Share	0.1/0.9	0.2/0.8	0.3/0.7	0.4/0.6	0.5
Settlement	$4,000	$7,000	$9,000	$10,000	$9,000

Table 4. Settlement from ISP to IBP

---

[9] 1.5 M bps * (3600 second * 4 peak hours * 30 days)

7 INTERNET ACCESS USER PRICE AND PEERING INCENTIVE

If we assume that each customer pays the fixed price per month for his accessing the Internet, by use of the cost and revenue function we can calculate how much a user price should be.

The following shows the procedure of calculation of user price P_1 and P_2 of ISP-1 and ISP-2 under the assumption of ISP-1's market share $0 < \alpha < 0.5$.

- Revenue of ISP-1 = P_1 * (Number of Subscribers)
  = $P_1 * (\alpha*5,000)$ ....(11)
- Cost of ISP-1 = (Settlement to IBP)
  = (No. of T-1 lines)*(T-1 transit price)
  = (No. of T-1 lines)* $1,000..................(12)
- Profit of ISP-1 = (11) – (12)
  = $P_1*(\alpha*5,000)$ - (No. of T-1 lines)* $1,000
  = $1,000*\{5*P_1*\alpha$ – No. of T-1 lines$\}$........(13)
- Equation (15) should be greater than '0'.
- User Price $P_1 >$ (No. of T-1)/(5*α) ....(14)

The same logic is applied to ISP-2.
- Revenue of ISP-2 = P_2 * (Number of Subscribers)
  = $P_2 * ((1-\alpha)*5,000)$ ....(15)
- Cost of ISP-2 = (Settlement to IBP)
  = (No. of T-1 lines)*(T-1 transit price)
  = (No. of T-1 lines)* $1,000 ....(16)
- Profit of ISP-2 = (15) – (16)
  = $P_2*((1-\alpha)*5,000)$ - (No. of T-1 lines)*$1,000
  = $1000*\{P_2*(1-\alpha)*5$ - (No. of T-1 lines)$\}$..(17)
- Equation (17) should be greater than '0'.
- User Price $(P_2) >$ (No. of T-1)/(5*(1-α)) ....(18)

From the equation (14) and (18) we can calculate the breakeven user price P_1 and P_2 for ISP-1 and ISP-2 to survive in the downstream market. The following table shows the user prices P_1 and P_2 according to the change of market share α from 0.1 to 0.9.

ISP-1	ISP-2	ISP-1		ISP-2		IBP	ISP-1	ISP-2
							Breakeven	Breakeven
α	1-α	T-1s	Settlement	T-1s	Settlement	Revenue	P1	P2
0.1	0.9	4	$4,000	4	$4,000	$8,000	$8	$0.89
0.2	0.8	7	$7,000	7	$7,000	$14,000	$7	$1.75
0.3	0.7	9	$9,000	9	$9,000	$18,000	$6	$2.57
0.4	0.6	10	$10,000	10	$10,000	$20,000	$5	$3.33
0.5	0.5	9	$9,000	9	$9,000	$18,000	$3.6	$3.6
0.6	0.4	10	$10,000	10	$10,000	$20,000	$3.33	$5
0.7	0.3	9	$9,000	9	$9,000	$18,000	$2.57	$6
0.8	0.2	7	$7,000	7	$7,000	$14,000	$1.75	$7
0.9	0.1	4	$4,000	4	$4,000	$8,000	$0.89	$8

Table 5. Breakeven User Price of ISPs

If the Internet access market user price is set to $2.5 per month, then we can divide two regions according to the market share. In the region of (0.3 ≤ α ≤ 0.7), both ISP-1 and ISP-2 will make a negative profit. Therefore, both ISPs have an incentive to make a peering arrangement. However, in the other region like (α < 0.3 or α > 0.7), the ISP with a larger market share has a positive profit and does not want to make a peering arrangement with the ISP with a smaller market share. This implies equality characteristic of peering. The following table shows each ISP's dominant strategy according to its market share.

ISP-1	ISP-2	ISP-1's	ISP-2's
α	1 - α	dominant Strategy	dominant Strategy
0.1	0.9	Peering	Transit
0.2	0.8	Peering	Transit
0.3	0.7	**Peering**	**Peering**
0.4	0.6	**Peering**	**Peering**
0.5	0.5	**Peering**	**Peering**
0.6	0.4	**Peering**	**Peering**
0.7	0.3	**Peering**	**Peering**
0.8	0.2	Transit	Peering
0.9	0.1	Transit	Peering

Table 6. ISP's Dominant Interconnection Strategy

In the peering arrangement, two ISPs can avoid expensive settlement payment to the IBP through bypassing an IBP's network and they can lease communication lines from bandwidth market to connect each other. The cost of the leased line can be divided by equally between two ISPs. They also know each other's inbound and outbound traffic volumes, because one ISP's inbound (outbound) traffic is the other ISP's outbound (inbound) traffic. Therefore Max $\{T_{1O}, T_{2O}\}$ criterion can also be applied in this case and there should be no settlement with each other. The only cost in this case is a half of the leased line to connect each ISP. The following graph shows relationship under the peering arrangement.

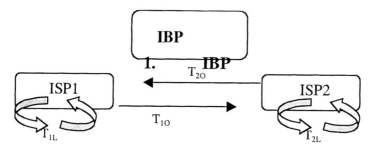

Figure 4. *Traffic Flow under Peering Arrangement*

To make a peering arrangement efficient, the cost of peering should be less than the lowest payment of settlement to the IBP. The minimum settlement from each ISP to the IBP in the region of $(0.3 < \alpha < 0.7)$ is $9,000 per month. The half of the price of the leased line from the peering arrangement should be less than $9,000 for the establishment of peering agreement.

For example, in the case of local peering, one T-1 lease line is assumed to be $300 per month[10], i.e., the half of monthly price of 10 T-1 lines is $1,500, which is far lower than the minimum settlement to the IBP. If we assume ISPs are located at east and west coast and they use T3 line instead of 10 T-1 lines to connect each other, the monthly lease price of T-3 line[11] is $6,000, the half of which is also lower than $9000.

8 NEGOTIATION

If the two ISPs make a peering agreement, the revenue of the IBP is '0'. Because any amount of settlement is better than '0', the IBP wants to try to negotiate with each IBP to lower settlement payment. There are two reasons for the IBP to make a negotiation with the two ISPs:

---

[10] www.bandwidthmarket.com (Pittsburgh local price, Jan. 2002) T-1($294)

[11] www.bandwidthmarket.com (LA – NYC, Jan. 2002), T-1($725), T-3($6,044)

1. If one of ISPs can not survive in the market, the relationship between upstream market and downstream market changes from (1-IBP vs. 2-ISPs) to (1-IBP vs. 1-ISP). In the monopoly-monopsony relationship, the IBP's negotiation power will reduce, that is not desirable to the IBP.
1. Expensive settlement makes the ISPs to raise their user access price, which can lead some of users out of market. If the number of users N decrease, outgoing traffic also decrease because outgoing traffic is dependent on the other ISP's market share. Finally the total market size may shrink and revenues of the IBP also shrink, that is also not desirable to the IBP.

9 APPLYING THIS MODEL TO THE REALITY

If we apply this model to the backbone market, the IBP in this model can be one of the top tiered IBPs[12] with a nationwide backbone network and ISP-1 and ISP-2 can be the second tiered IBPs which need Internet connectivity to those top tiered IBPs. In 1997, starting with UUNET, top tiered backbone providers announced not to peer with smaller backbone providers and content specialized backbone providers. (Kende, 2000) Five top tiered Internet backbone providers control nearly 80% of the Internet backbone (Weinberg, 2000), which means the Big Five's market power is enough to control over the whole Internet backbone market. The peering arrangement has been set among the Big Five, but everyone else must pay the price for passing traffic over Big Five's networks, which means the rest of IBPs have to make a transit agreement with Big Five. Because the terms and conditions of the peering agreement are usually not open, called 'Non Disclosure Agreement' and the other alternative to connect to the Internet is through Public Network Access Points (NAPs), which have been suffered by congestion. In reality, the bargaining power of Big Five would be greater than expected. According to the number of backbone providers, the market could be classified as a competitive market, but according to the Big Five's behaviors, the market could be closer to oligopoly. Their settlement decision is only based on the line capacity between two backbone providers regardless of traffic volume. If the industry uses the maximum {inbound traffic volume, outbound traffic volume}, it would be more effective and efficient to motivate peering arrangement with smaller backbone providers. If the second tiered backbone providers made a multilateral private peering

---

[12] According to Michael Kende's Digital Handshake, top tiered IBPs are Worldcom (UUNET), AT&T, Cable and Wireless, Genuity, and PSINet. According to Neil Weinberg's Backbone Bullies, top tiered IBPs are Worldcom, Genuity, AT&T, Sprint, and Cable and Wireless.

or to notify a system administrator of a problem. Voice could be in the form of an audio attachment, like VocalTec's Internet Voice Mail 3.0 (8) or making real time voice contact over an IP connection through the desktop interface. Committees within the Internet Engineering Task Force (IETF) are working on specifications, likely as extensions of the Multipurpose Internet Mail Extensions (MIME) format, for all these features. Internet messaging will eventually catch up to proprietary systems' reliability and security, but parity is at least two years away (8).

In proprietary systems like in Novell 5 or Lotus Notes 5, a universal mailbox is already available. User can perform document management task, send a fax, schedule appointments, maintain a task list, or perform other functions through the mailbox and are accessible to people who are travelling, from a remote location.

6. REFERENCES

[1] E. Bowes: The Three-Tier Shuffle. Database Design, Volume 10, Number 9, 1997.
[2] P. Wayner: Inside the NC. Byte, November 1996
[3] T. Halfhill: Cheaper Computing. Byte, April – May 1997.
[4] B. Pierce: Combining Java and CGI Scripts, communicating between client and server. WEB Design, Volume 2, Issue 9. 1997.
[5] E.H. Harold: Java Network Programming. O'Reilly and Associates, 1997.
[6] S. Spainhour and V. Quercia: Webmaster in a Nutshell. O'Reilly and Associates, 1996.
[7] S. Gundavaram: CGI Programming on the WWW. O'Reilly and Associates. 1996.
[8] M. Nadeau: Your E-Mail is Obsolete, Byte, February 1997.